METALLURGY – ADVANCES IN MATERIALS AND PROCESSES

Edited by **Yogiraj Pardhi**

Metallurgy – Advances in Materials and Processes
http://dx.doi.org/10.5772/2852
Edited by Yogiraj Pardhi

Contributors

William A. Brantley, Satish B. Alapati, Hiroto Kitaguchi, Jakub Siegel, Ondřej Kvítek, Zdeňka Kolská, Petr Slepička, Václav Švorčík, Ji Fan, Chuan Seng Tan, Mohammad Hosein Bina, Ion Pencea, Dejan Tanikić, Vladimir Despotović

Published by InTech

Janeza Trdine 9, 51000 Rijeka, Croatia

Copyright © 2012 InTech

Notice

Publishing Process Manager Marijan Polic
Technical Editor InTech DTP team
Cover InTech Design team

Additional hard copies can be obtained from orders@intechopen.com

Metallurgy – Advances in Materials and Processes, Edited by Yogiraj Pardhi
p. cm.
ISBN 978-953-51-0736-1

For EU product safety concerns:
IN TECH d.o.o., Prolaz Marije Krucifikse Kozulić 3, 51000 Rijeka, Croatia, info@intechopen.com or visit our website at intechopen.com.

Meet the editor

Dr Pardhi is currently working in the School of Metallurgy and Materials at the University of Birmingham, United Kingdom. He has expertise in materials characterisation, process development and mechanical property characterisation of high performance materials. Dr. Pardhi has done undergraduate degree in Mechanical Engineering at the Nagpur University, India. He finished his Master of Technology in Industrial Metallurgy from the Indian Institute of Technology Roorkee, India. During his masters degree he was awarded the DAAD (German Academic Exchange Service) fellowship for duration of 1 year to carry out his dissertation at the Technical University of Munich, Germany. He went on to do his PhD in Metallurgy and Materials, at the University of Birmingham. He was awarded the prestigious ORSAS-2006 fellowship by Universities UK during his doctoral research.

Contents

Preface

In recent decades scientists and engineers around the globe have been responding to the requirement of high performance materials through innovative material research and engineering. The ever increasing demand on quality and reliability has resulted in some dazzling technological achievements in the area of advanced materials and manufacturing. A number of new tools and techniques have been developed along with extending current capabilities from performance perspective.

The purpose of this book is to bring together significant findings of leading experts, in developing and improving the technology that supports advanced materials and process development. The contributions made by researchers in these fields are immensely valuable. From gold nano-structures to advanced superalloys, this book covers investigations involving modern computer based approaches as well as traditional experimental techniques. Some of the techniques described in this book include, artificial intelligence based approaches to metal cutting, multi-conventional mathematical model based technique to the measure uncertainties in metallurgical tests and low temperature bonding technology for 3D integration. Selected articles include research output on advances made in materials that are used not only in complex structures such as aeroplanes but also in clinical treatments. There are chapters that present research on Ni based superalloys, gold nanostructures, heat resistance steels and advanced dental alloys. Individual texts include introduction to the topics presented, illustrative procedures used in the investigations, results of study undertaken and qualitative discussion, based on the findings, with a summary at the end.

This book is formulated with chapters which describe the most recent work in materials and process development at the time of publication. It is envisaged that it will promote knowledge transfer across the materials society including university students, engineers and scientists to built further understanding of the subject. It is assumed that the reader has elementary knowledge of the materials and processes described in this book. In cases where details are required, appropriate references will assist the further understanding. It is hoped that, the reader will find the work presented exciting, challenging and valuable as the original investigators intended.

The editor would like to thank all contributing authors for their hard work and patience throughout the publication process. Acknowledgement is also due to InTech open science and the publication process manager for their excellent cooperation during review and editing stages.

Dr. Yogiraj Pardhi
School of Metallurgy and Materials, University of Birmingham,
United Kingdom

Chapter 1

Heat Treatment of Dental Alloys: A Review

William A. Brantley and Satish B. Alapati

Additional information is available at the end of the chapter

http://dx.doi.org/10.5772/52398

1. Introduction

Metallic materials have widespread use in dentistry for clinical treatment and restoration of teeth. Major areas of usage are: (1) restorative dentistry and prosthodontics (dental amalgam and gold alloy restorations for single teeth, metallic restorations for multiple teeth, including metal-ceramic restorations, removable partial denture frameworks, and dental implants), (2) orthodontics (wires which provide the biomechanical force for tooth movement), and (3) endodontics (rotary and hand instruments for treatment of root canals). Heat treatment of the metal can be performed by the manufacturer, dental laboratory, or dentist to alter properties intentionally and improve clinical performance. Heat treatment of the metal also occurs during the normal sequence of preparing a metal-ceramic restoration, when dental porcelain is bonded to the underlying alloy substrate. Moreover, intraoral heat treatment of some metallic restorations occurs over long periods of time. There is an enormous scientific literature on the heat treatment of metals for dentistry. A search of the biomedical literature in May 2012, using PubMed [http://www.ncbi.nlm.nih.gov/pubmed/] revealed nearly 450 articles on heat treatment of dental alloys. The purpose of this chapter is to provide a review of the heat treatment of metallic dental materials in the foregoing important areas, describing the important property changes, with a focus on the underlying metallurgical principles.

2. Restorative dentistry and prosthodontics

2.1. Dental amalgams

Dental amalgams are prepared in the dental office by mixing particles of a silver-tin-copper alloy for dental amalgam that may contain other trace metals with liquid mercury. The initially mixed (termed triturated) material is in a moldable condition and is placed (termed condensed) directly by the dentist into the prepared tooth cavity, where it undergoes a setting process that produces multiple phases and can require up to one day for near completion. Extensive information about the several different types of dental amalgams are provided in textbooks on dental materials [1,2]. Particles of the alloy for dental amalgam are manufactured by either lathe-cutting a cast ingot or directing the molten alloy through a special nozzle. Both the machining of the lathe-cut particles and the rapid solidification of the spherical particles create residual stress. In addition, the microstructure of the solidified silver-tin-copper alloy has substantial microsegregation. Consequently, manufacturers of the alloy powder for dental amalgam perform a proprietary heat treatment to relieve residual stresses and obtain a more homogeneous microstructure. This heat treatment is of considerable practical importance since it affects the setting time of the dental amalgam after the powder is mixed with mercury. Subsequently, the dental amalgam restorations undergo intraoral aging, which can be regarded as heat treatment, and detailed information about the microstructural phase changes for prolonged intraoral time periods has been obtained from clinically retrieved dental amalgam restorations [3].

2.2. Gold alloys for all-metal restorations

Gold alloys are principally used for all-metal restorations (inlays, crowns and onlays) in single posterior teeth. These alloys are cast by a precision investment process, and the restorations are cemented by the dentist into the prepared tooth cavity. The original gold casting alloys contained over approximately 70 wt.% gold, but the very high price of gold has led to the development of alloys that contain approximately 50 wt.% gold. These alloys also contain silver, copper, platinum, palladium, zinc, and other trace elements, including iridium for grain refinement. Information about the dental casting process and the gold alloys is available in dental materials textbooks [1,2]. Detailed compositions and mechanical properties of specific alloys are available on the website of the major manufacturers. Another valuable reference is the current ISO Standard on metallic materials for fixed and removable dental appliances [4], which stipulates mechanical property requirements. In the normal dental laboratory procedure, gold castings for all-metal restorations are water-quenched after solidification, following loss of the red heat appearance for the sprue. This results in formation of a disordered substitutional solid solution and leaves the alloy in the soft condition, which is preferable since adjustments are more easily made on the restoration by the dental laboratory or dentist. The gold alloy casting can also be placed in the soft condition by heating at 700°C for 15 minutes and water-quenching. The quenched gold casting may be placed in the hard condition by heat treatment at 350°C for 15 minutes and air-cooling. This heat treatment results in formation of ordered AuCu or $AuCu_3$ regions in the disordered matrix of the high-gold or lower-gold alloys, respectively. Examples of changes in clinically important mechanical properties from heat treatment are shown in Table 1 for two gold alloys, where (S) and (H) represent the soft and hard conditions.

In practice, dental laboratories do not perform heat treatments on the cast gold restorations because of the time involved. However, it appears to be fortunate that the gold alloys that contain sufficient copper to undergo ordering will undergo age hardening in the mouth.

Figure 1 compares the intraoral aging behavior of a traditional high-gold dental alloy (Type IV) and a special gold alloy containing gallium (AuCu-3wt%Ga) [5].

Alloy	Vickers Hardness		0.2% Offset Yield Strength		Percentage Elongation	
Firmilay (74.5% Au)	121 (S)	182 (H)	207 MPa (S)	276 MPa (H)	39% (S)	19% (H)
Midas (46% Au)	135 (S)	230 (H)	345 MPa (S)	579 MPa (H)	30% (S)	13% (H)

Table 1. Summary of property changes resulting from heat treatment of two gold alloys for all-metal restorations. [http://www.jelenko.com/, accessed August 15, 2012]

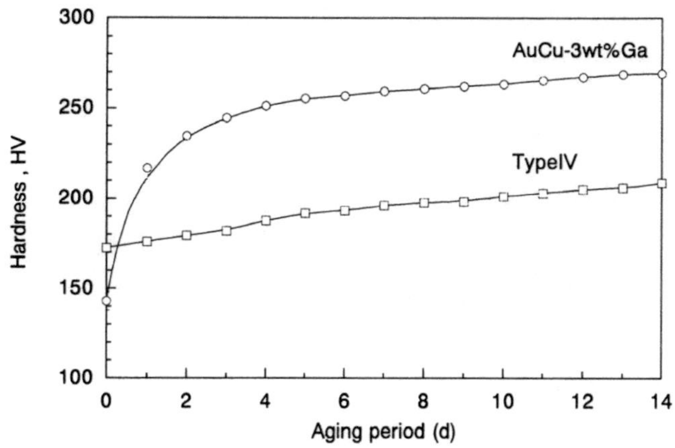

Figure 1. Comparison of the two-week aging behavior at 37°C for a high-gold dental alloy and a dental gold alloy containing gallium that was designed to undergo intraoral aging. From [5] and reproduced with permission.

2.3. Alloys for fixed prosthodontics (metal-ceramic restorations)

Metal-ceramic restorations are in widespread clinical use for restorative and prosthetic dentistry, and are employed for single-tooth restorations and for restorations involving multiple adjacent teeth (fixed prostheses or crown-and-bridgework). An alloy is cast using the precision investment procedure in dental laboratories to fit accurately to the prepared tooth or teeth, and to form a substrate (termed the coping) for the porcelain. After an initial oxidation step that forms a native oxide on the metal surface, one or two layers of opaque porcelain are bonded to the metal, followed by the application of a layer of body porcelain and a surface glaze [1,2]. In order to have a strong bond between the porcelain and metal, which is essential for clinical longevity of the metal-ceramic restoration, the coefficients of thermal contraction for the metal and porcelain must be closely matched, and a difference not exceeding 0.5 ppm/°C is generally desired. Mechanical property requirements for the alloys are stipulated in ANSI/ADA Specification No. 38 (ISO 9693) [6], and the minimum value of 250 MPa for the 0.2% offset yield strength is important, since the thin coping must withstand intraoral forces without undergoing permanent deformation. The metal-ceramic bond strength (termed the bond compatibility index) is measured with a three-point bending test that uses thin cast alloy strip specimens having a centrally located area of sintered porcelain, and a minimum bond strength (shear stress) of 25 MPa is stipulated.

Both noble and base metal alloys are used for bonding to dental porcelain. The current American Dental Association classification has four alloy groups for fixed prosthodontics [7]: (1) high-noble (gold-platinum-palladium, gold-palladium-silver and gold-palladium); (2) noble (palladium-silver, palladium-copper-gallium, and palladium-gallium); (3) predominantly base metal (nickel-chromium and cobalt-chromium); (4) titanium and titanium alloys. Information about these alloys for metal-ceramic bonding is summarized in a textbook on fixed prosthodontics [8]. The principal mechanisms for metal-ceramic bonding are (a) mechanical interlocking from the initially viscous porcelain at the elevated sintering temperatures flowing into microirregularities on the air-abraded cast metal surface and (b) chemical bonding associated with an interfacial oxide layer between the metal and ceramic. These two mechanisms are evident from photomicrographs, found in numerous references [8], of the fracture surfaces for metal-ceramic specimens prepared from a wide variety of dental alloys. This native oxide forms on the cast alloy during the initial oxidation firing step in the dental porcelain furnace. Noble alloys for bonding to dental porcelain contain small amounts of secondary elements, such as tin, indium and iron, which form the native oxide and also increase the alloy strength. However, Mackert et al [9] found that during initial oxidation heat treatment, metallic Pd-Ag nodules formed on the surface of a palladium-silver alloy for metal-ceramic restorations and only internal oxidation occurred for the tin and indium present in the alloy composition. They concluded that porcelain bonding arose predominantly from mechanical interlocking with the nodules. Internal oxidation has also been reported for high-gold [10] and high-palladium [11] alloys for bonding to porcelain, but both alloy types also formed surface oxides [10,12].

The initial oxidation step and subsequent sintering (also termed baking or firing) of the dental porcelain layers causes the alloy to experience substantial heat-treatment effects. Under normal dental laboratory conditions, the porcelain firing sequence is performed rapidly. For example, in one study heating of high-palladium alloys in the dental porcelain furnace was performed at approximately 30°C/min over a temperature range from 650°C to above 900°C, and the total heating time for the several firing cycles at these elevated temperatures was about 45 minutes [11]. Studies [13-15] have shown that the as-cast microstructures of noble metal alloys for bonding to porcelain are highly inhomogeneous in the initial as-cast condition, presumably from substantial elemental microsegregation that occurs during the rapid solidification involved with casting into much cooler investment [1,2]. After simulation of the dental porcelain firing sequence, the noble metal alloy microstructures become substantially homogeneous, and there are accompanying changes in the mechanical properties, as shown in Table 2.

Peaks in Vickers hardness for heat treatments at temperatures that span the porcelain-firing temperature range indicate that influential precipitation processes can occur in some noble alloys for fixed prosthodontics [13,16]. For the gold-palladium-silver alloy in Table 1, heating an as-cast specimen to 980°C caused a pronounced decrease in Vickers hardness, and subsequent heat treatments at temperatures from 200° to 980°C revealed a pronounced peak in Vickers hardness at approximately 760°C. The absence of substantial changes in Vickers hardness for similar heat treatments of the gold-palladium alloy in Table 2 arises from differences in the precipitates that form in the two complex alloy compositions. Figure 2 presents the age hardening behavior of a palladium-silver alloy, where specimens were subjected to isothermal annealing for 30 minute time periods at temperatures from 400°C to 900°C that span the range for the porcelain firing cycles [16]. Bulk values of Vickers hardness were obtained with 1 kg loads, and 25 g loads were used to obtain hardness values for specific microstructural regions. In contrast, research suggests that microstructures of popular nickel-chromium base metal alloys used with dental porcelain are not changed substantially during dental laboratory processing [17].

Alloy Type	Vickers Hardness		0.2% Offset Yield Strength		Percentage Elongation	
Au-Pd-Ag (Neydium)	199 (C)	218 (P)	420 MPa (C)	490 MPa (F)	6% (C)	8% (F)
Au-Pd (Olympia)	213 (C)	225 (P)	500 MPa (C)	540 MPa (F)	13% (C)	20% (F)

Table 2. Mechanical properties for two noble metal alloy types used with dental porcelain, comparing the as-cast condition (C) and simulated porcelain firing heat treatment (F) [13].

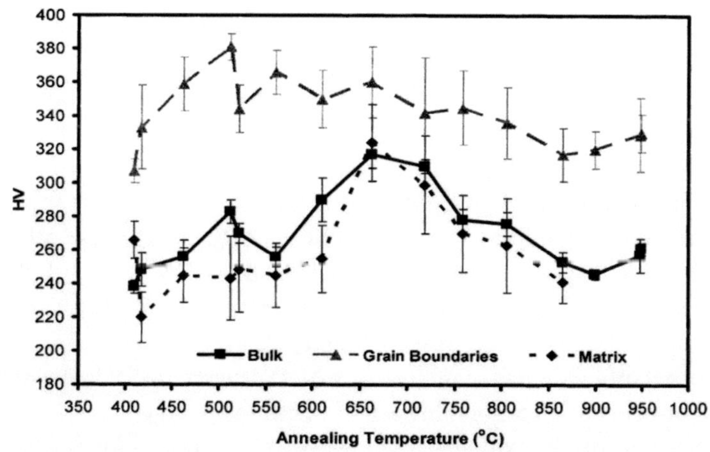

Figure 2. Annealing behavior of a palladium-silver alloy for fixed prosthodontics, showing changes in Vickers hardness for a heat treatment temperature range that spans the porcelain firing cycles. Reproduced from [16] with permission.

2.4. Alloys for removable prosthodontics

Base metal casting alloys (nickel-chromium, cobalt-chromium and cobalt-chromium-nickel) are popular for fabricating the metallic frameworks for removable partial dentures because of their lower cost [1,2]. Once an active area of dental metallurgy research, studies have found that these alloys have dendritic microstructures in the as-cast condition, because of the absence of suitable grain-refining elements, and that heat treatment is ineffective for producing improved mechanical properties [18]. A more recent publication shows the dendritic microstructures of some current alloys and their mechanical properties [19].

Removable partial denture frameworks have clasps that engage the teeth. These clasps can be cast as part of the entire framework, or alternatively wire clasps can be joined to the cast framework in the dental laboratory [1,2]. Both noble metal and base metal wires for clasps are available [20]. Because of their superior strength compared to the cast base metal alloys, wire clasps with smaller cross-section dimensions can be used with the frameworks, but caution is required during joining in the dental laboratory to avoid overheating that will cause loss of the wrought microstructure. Wire clasps are used in the as-received condition; heat treatment is not recommended before joining to the framework.

2.5. Dental implant alloys

Dental implants in current widespread clinical use are manufactured from CP (commercially pure) titanium or Ti-6Al-4V, and some implants have a thin bioceramic surface coating (typically hydroxyapatite, the principal inorganic constituent of bone and tooth structure). Proprietary heat treatments [21] are performed on Ti-6Al-4V by manufacturers to obtain optimum microstructures for the implants; minimal information is currently available about these microstructures in the dental scientific literature.

Recently, there has been considerable research interest in the development of new titanium implant alloys for orthopedic applications that have improved biocompatibility compared to the Ti-6Al-4V alloy in widespread current use. There is particular interest in the beta-titanium alloys which have lower elastic modulus than Ti-6Al-4V to minimize stress shielding and subsequent loss of the surrounding bone which has a much lower elastic modulus. Stress shielding does not seem to be of concern for dental implants, presumably because of the threaded designs. Biocompatible titanium-niobium-zirconium beta alloys have been investigated, and oxide nanotubes can be grown on the alloy surface by an anodization technique, and subsequent heat treatment can be employed to modify the structure of the nanotubes [22]. In another exciting research area, titanium oxide nanowires have been recently grown on both CP titanium and Ti-6Al-4V using special elevated-temperature oxidation heat treatments in an argon atmosphere with low oxygen concentrations [23]. Both of these special types of surface oxide layers may prove to be useful for dental and orthopedic implants, but future testing in animals will be needed to examine their efficacy.

3. Orthodontics

3.1. Background

Orthodontic wires engaged in brackets that are bonded to teeth, after being deformed elastically during initial placement, provide the biomechanical force for tooth movement during unloading. There are four wire types in current clinical practice: stainless steel, cobalt-chromium, beta-titanium and nickel-titanium [24]. The clinically important mechanical properties are (a) elastic modulus, which is proportional to the biomechanical force when wires of similar dimensions are compared; (b) springback, which is generally expressed as the quotient of yield strength and elastic modulus (YS/E), and represents the approximate strain at the end of the clinically important elastic range; and (c) modulus of resilience, expressed as $YS^2/2E$ and representing the spring energy available for tooth movement. (The permanent deformation portion of orthodontic wire activation is ineffective for tooth movement.) Round orthodontic wires are manufactured by a proprietary drawing sequence that involves several stages with intermediate annealing heat treatments. Rectangular orthodontic wires are manufactured by a rolling process utilizing a Turk's head apparatus. The wire drawing process with the heat treatments greatly affects mechanical properties.

3.2. Stainless steel orthodontic wires

A recent study that investigated stainless steel wires used in orthodontic practice found that most products were AISI Type 304 and that AISI Type 316L (low carbon) and nickel-free ASTM Type F2229 were also available [25]. While standard physical metallurgy textbooks consider the elastic modulus to be a structure-insensitive property, research has shown that the permanent deformation and heat treatments involved with the wire drawing process can substantially affect the elastic modulus of stainless steel orthodontic wires [26,27]. X-ray diffraction has revealed that conventional orthodontic wires manufactured from AISI Types 302 and 304, while predominantly austenitic structure, can contain the α' martensitic phase, depending upon the carbon content and temperatures involved with the processing [28]. The presence of this martensitic phase accounts for the reduction in elastic modulus for some conventional stainless steel orthodontic wires. In addition, when fabricating complex stainless steel appliances, it is recommended that orthodontists perform a stress-relief heat treatment to prevent fracture during manipulation; a heating time up to 15 minutes and a temperature range of 300° to 500°C appears to be acceptable [29-31]. Heating austenitic stainless steel to temperatures between 400° and 900°C can result in chromium carbide precipitation at grain boundaries and cause the alloy to become susceptible to intergranular corrosion, and heating of austenitic stainless steel wires above 650°C should not be done because loss of the wrought microstructure causes degradation of mechanical properties.

3.3. Cobalt-chromium orthodontic wires

The cobalt-chromium orthodontic wire (Elgiloy) marketed by Rocky Mountain Orthodontics (Denver, CO, USA) contains 40% Co, 20% Cr, 15.81% Fe, 15% Ni, 7% Mo, 2% Mn, 0.15% C carbon and 0.04% Be beryllium (https://www.rmortho.com/, accessed August 15, 2012). Four different tempers (spring quality) are available, and the soft Blue temper is favored by many orthodontists because the wire is easily manipulated in the as-received condition, and then heat treated to increase the yield strength and modulus of resilience. Heat treatment (not recommended for the most resilient temper) is conveniently performed with the electrical resistance welding apparatus commonly used in orthodontic practice, and the manufacturer provides a special paste that indicates when the heat treatment is complete. Alternatively, furnace heat treatment performed at 480°C for 5 minutes has been found to give satisfactory results [32]. An extensive study employing furnace heat treatment (480°C for 10 minutes) for three tempers and numerous sizes of the Elgiloy wires observed increases of 10% – 20% in elastic modulus and 10% – 20% in 0.1% offset yield strength, which resulted in substantial improvement of the modulus of resilience [27]. These changes in mechanical properties arise from complex precipitation processes during heat treatment that are not understood. Many other companies now market cobalt-chromium orthodontic wires, but studies of their mechanical properties and the results of heat treatment have not been reported.

3.4. Beta-titanium and other titanium-based orthodontic wires

Beta-titanium orthodontic wires have the advantages of: (a) known biocompatibility from the absence of nickel in the alloy composition; (b) lower elastic modulus than stainless steel and cobalt-chromium wires, which provides more desirable lower orthodontic force for tooth movement; (c) higher springback than stainless steel and cobalt-chromium wires, which is desirable for the archwire to have greater elastic range; and (d) high formability and weldability, which are needed for fabrication of certain appliances [24]. A recent study [25] of commercially available titanium-based orthodontic wires revealed that most products are Beta III alloys [21] containing approximately 11.5 Mo, 6 Zr, and 4.5 Sn, similar to the original beta-titanium wire introduced to orthodontics [33,34]. Beta C [21] and Ti-45Nb beta-titanium and Ti-6Al-4V (alpha-beta) wire products are also available [25]. Heat treatment is not performed by the orthodontist on these wires, but care with the wire drawing and intermediate heat treatments by the manufacturer are essential for obtaining the desired mechanical properties. These processes must be conducted under well-controlled conditions because of the highly reactive nature of titanium.

3.5. Nickel-titanium orthodontic wires

Following the pioneering work of Andreasen and his colleagues [35,36], near-equiatomic nickel-titanium (NiTi) wire was introduced to orthodontics by the Unitek Corporation (now 3M Unitek) [37]. This wire had the advantages of a much lower elastic modulus than the stainless steel and cobalt-chromium wires available at the time and a very large elastic range. The clinical disadvantage is that substantial permanent deformation of this wire is not possible to obtain certain orthodontic appliances that can be fabricated with the three preceding, highly formable, alloys. The original nickel-titanium wire had a work-hardened martensitic structure and did not exhibit the superelastic behavior (termed pseudoelasticity in engineering materials science) or the true shape memory characteristics displayed by subsequently introduced NiTi wires [1,38-41]. These nickel-titanium wires have been a very active area of research.

The mechanical properties of the nickel-titanium orthodontic wires are determined by the proportions and character of three microstructural phases: (a) austenite, which occurs under conditions of high temperature and low stress; (b) martensite, which occurs under conditions of low temperature and high stress; and (c) R-phase, which forms as an intermediate phase during the transformation between martensite and austenite. Very careful control of the wire processing and associated heat treatments, along with precise compositional control, by the manufacturer are needed to produce nickel-titanium wires with the desired superelastic, nonsuperelastic, or shape memory character [42,43].

Heat treatments have been exploited by manufacturers to control the orthodontic force ranges produced by nickel-titanium archwires [39]. Heat treatment temperatures have ranged from 400° to 600°C with times from 5 minutes to 2 hours [39,40]. Effects of heat treatment on cantilever bending plots for two sizes of a round superelastic nickel-titanium wire are presented in Figure 3 [40].

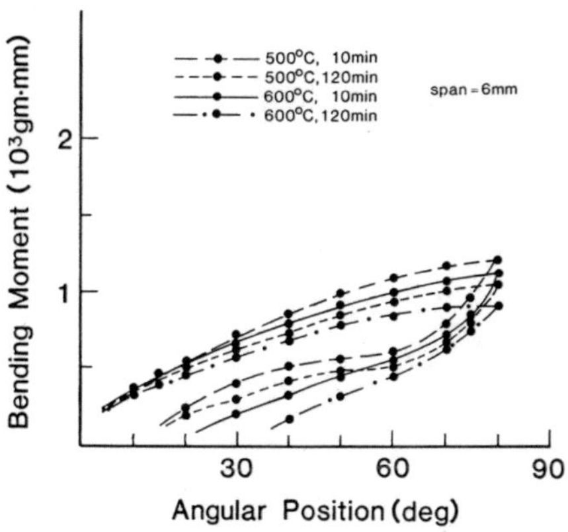

Figure 3. Effects of heat treatments on cantilever bending plots for 6 mm test spans of a superelastic nickel-titanium orthodontic wire. Reproduced from [40] with permission.

Loss of superelastic behavior occurs for the 2 hour heat treatment at 600°C, evidenced by the large decrease in springback (difference between the original deflection of 80 degrees and the final angular position on unloading). Heat treatment at 500°C for 10 minutes had minimal effect, while heat treatment for 2 hours caused a decrease in the average superelastic bending moment during the unloading region of clinical importance. Bending properties for nonsuperelastic wires were only slightly affected by these heat treatments. In addition to the use of furnace heat treatment, electrical resistance heat treatment [44] has also been exploited by one manufacturer to produce archwires where the level of biomechanical force varies with position along the wire [24].

Microstructural phases at varying temperatures in nickel-titanium orthodontic wires and their transformations are conveniently studied by differential scanning calorimetry (DSC) [45]. Temperature-modulated DSC provides greater insight into the transformations than conventional DSC [46]. Figures 4 and 5 present temperature-modulated DSC heating curves for shape memory and superelastic nickel-titanium orthodontic wires, respectively. The transformations involving austenite (A), martensite (M) and R-phase (R) are labeled. The austenite-finish (A_f) temperature for completion of the transformation from martensite to austenite on heating is determined by the intersection with the adjacent baseline of a tangent line to the peak for the final transformation to austenite [47].

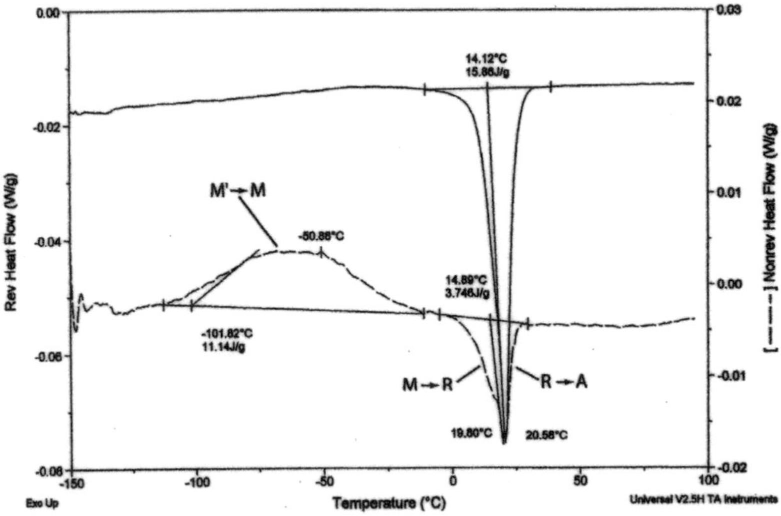

Figure 4. Heating temperature-modulated DSC plot for a shape memory nickel-titanium orthodontic wire. Reproduced from [46] with permission.

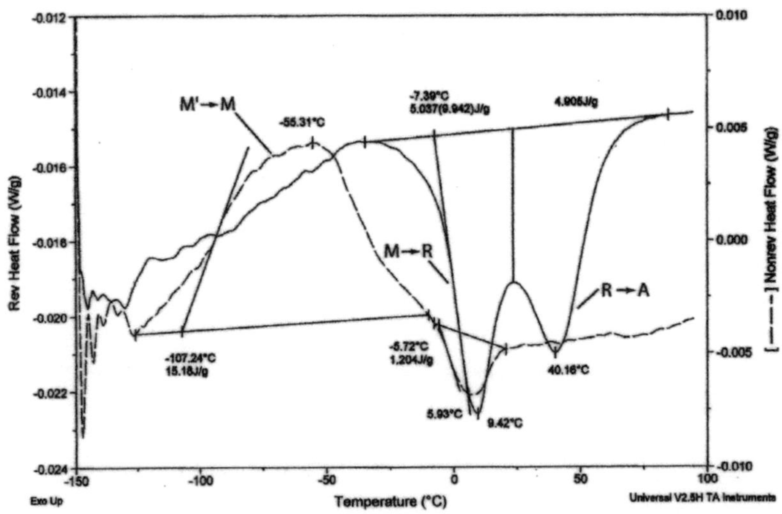

Figure 5. Heating temperature-modulated DSC plot for a superelastic nickel-titanium orthodontic wire. Reproduced from [46] with permission.

The A_f temperature is below body temperature (37°C) for nickel-titanium wires that exhibit shape memory in the oral environment. The superelastic nickel-titanium wires have A_f temperatures that are greater than mouth temperature and have more widely separated peaks for the successive transformations from M → R and R → A. The nonsuperelastic wires have much weaker transformations (lower values of enthalpy [ΔH]) and A_f temperatures that are also greater than mouth temperature [45]. Examination of x-ray diffraction patterns for nickel-titanium orthodontic wires revealed the effects of heat treatment on the M_s temperature for the start of the cooling transformation to martensite as well as the occurrence of stress relief and perhaps some recrystallization [24,48].

Transformation of a low temperature martensite phase (M′) to the higher temperature form of martensite (M), shown in Figures 5 and 6, is readily detected as a large exothermic peak on the nonreversing heat flow curves from temperature-modulated DSC. Transmission electron microscopy has revealed that this transformation arise from low-temperature twinning within the martensite structure [49].

4. Endodontics

4.1. Stainless steel instruments

Traditionally, endodontic treatment was performed with stainless steel hand files and reamers to remove the injured or diseased dental pulp from the root canals of teeth. While conventional elevated-temperature heat treatment is not recommended for these instruments, they are subjected to sterilization procedures before being using again with a different patient. One study found that dry heat sterilization (180°C for 2 hours) and autoclave sterilization (220 kPa pressure and 136°C for 10 minutes) slightly decreased the flexibility and resistance to torsional fracture of the instruments but they still satisfied the requirements for minimum angular deflection in the ISO standard [50]. Further research is needed to gain insight into the metallurgical origins of the property changes.

4.2. Nickel-titanium instruments

Following the pioneering work of Walia et al that introduced the nickel-titanium hand file to the endodontics profession [51], engine-driven rotary instruments were introduced that enable rapid instrumentation of root canals. These instruments are in widespread clinical use, and research on the nickel-titanium files has been a highly intensive area of research.

The major mechanical property of the equiatomic nickel-titanium alloy that led to replacement of the traditional austenitic stainless steel files was the much lower elastic modulus of NiTi, which enabled curved root canals to be negotiated with facility. An excellent review article [52] describes the manufacturing process for the nickel-titanium files, which are generally machined from starting wire blanks. The conventional nickel-titanium rotary instruments have been fabricated from superelastic nickel-titanium blanks.

Defects caused by the machining process and metallurgical flaws in the starting blanks, along with inadvertent overloading by the clinician, can result in fracture of the file within the root canal, which causes considerable patient anguish since the broken fragments often cannot be easily retrieved [53,54].

A recent study investigated the effect of heat treatment on conventional nickel-titanium rotary instruments, using temperature-modulated DSC and Micro-X-ray diffraction [55]. Results are shown in Figure 6 (a) – (d) for heat treatment at temperatures from 400° to 800°C in a flowing nitrogen atmosphere.

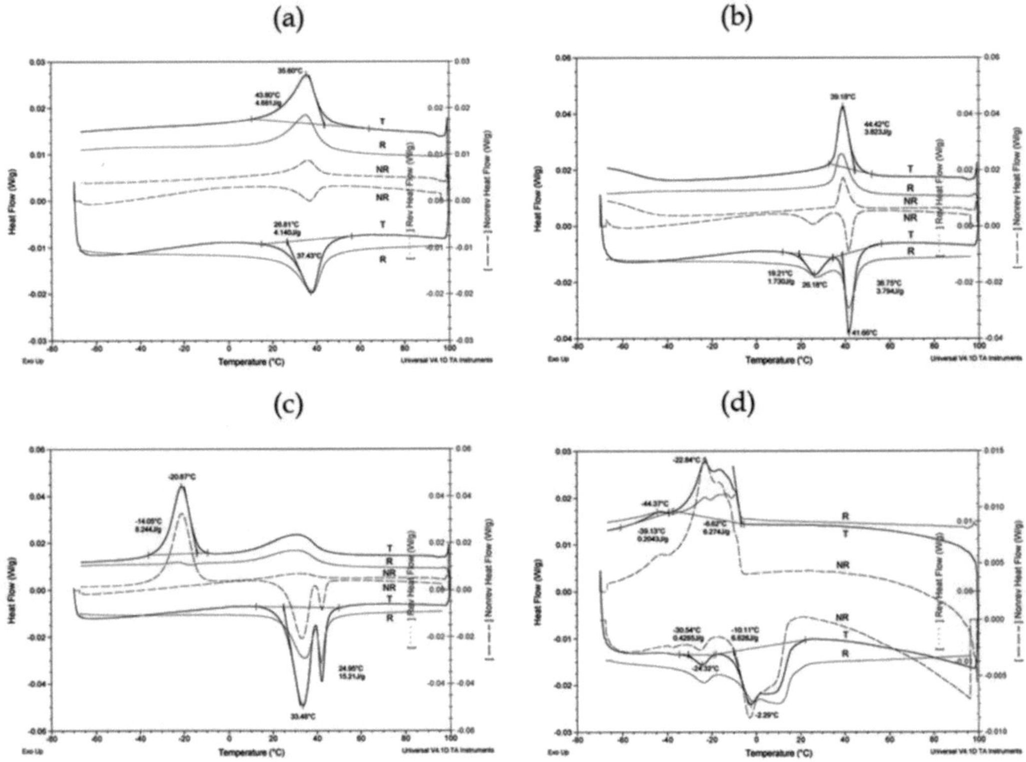

Figure 6. Temperature-modulated DSC reversing (R), nonreversing (NR) and total (T) heat flow curves for specimens from conventional rotary endodontic instruments after heat treatment in flowing nitrogen for 15 minutes at (a) 400°, (b) 500°, (c) 600° and (d) 850°C. From [55] and reproduced with permission.

Heat treatment between 400° and 600°C increased the A_f temperature for as-received conventional NiTi rotary instruments to approximately 45° – 50°C, and the transformations between martensite and austenite were changed to a more reversing character than nonreversing character [55]. Heat treatment in a nitrogen atmosphere might lead to a harder surface from the formation of nitrides [56], which is beneficial for cutting efficiency of the rotary instrument. This research suggested that heat treatment at temperatures near 500°C in a nitrogen atmosphere might yield the optimum microstructure and mechanical properties, with improved resistance to deformation and fracture for conventional NiTi rotary instruments. Heat treatment at temperatures exceeding 600°C should not be performed, since the superelastic behavior is lost along with potential degradation of the wrought microstructure [24]. Another study has reported that heat treatment at 430° and 440°C greatly improved the fatigue resistance of one conventional rotary instrument product [57].

New nickel-titanium rotary instruments have been marketed, for which the wire blanks were improved by special proprietary processing techniques, including heat treatment. The first notable example was M-Wire, named for its stable martensitic structure [58]. Previous conventional rotary instruments were fabricated from superelastic wire blanks with evident transformable austenite detected by conventional DSC [59]. However, when the conventional instruments were cooled far below room temperature to attain the fully martensite condition, the enthalpy changes for transformations from martensite to austenite were far below those for superelastic orthodontic wires [44,45], indicating that these instruments contain a substantial proportion of stable martensite in their microstructures.

Two different batches of M-Wire (termed Type 1 and Type 2), with unknown differences in proprietary processing, were obtained for characterization by temperature-modulated DSC and Micro-X-ray diffraction [58]. Figure 7 shows the differences in the temperature-modulated DSC plots for (a) conventional superelastic wire and (b) Type 1 M-Wire.

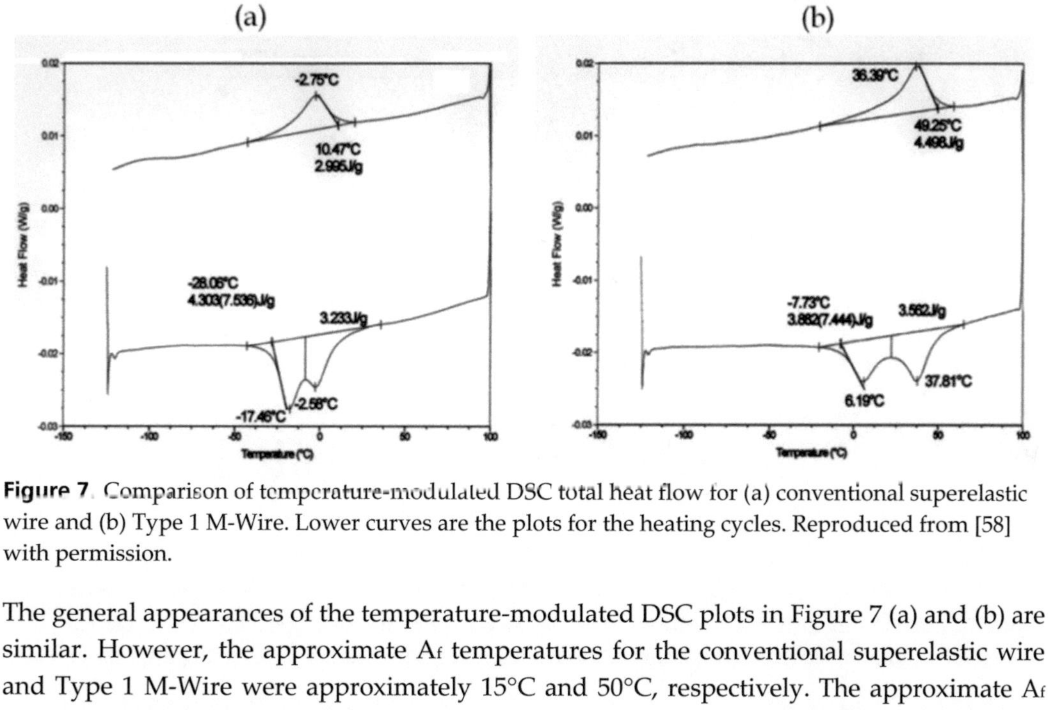

Figure 7. Comparison of temperature-modulated DSC total heat flow for (a) conventional superelastic wire and (b) Type 1 M-Wire. Lower curves are the plots for the heating cycles. Reproduced from [58] with permission.

The general appearances of the temperature-modulated DSC plots in Figure 7 (a) and (b) are similar. However, the approximate A_f temperatures for the conventional superelastic wire and Type 1 M-Wire were approximately 15°C and 50°C, respectively. The approximate A_f temperature for the Type 2 M-Wire was 45°C. The proportions of the different NiTi phases were quite different for Type 1 and Type 2 M-Wire, as shown in Figure 8.

The Micro-X-ray diffraction pattern indicated that Type 1 M-Wire had an austenitic structure, and the Micro-X-ray diffraction pattern from the conventional superelastic wire was similar. In contrast, the Micro-X-ray diffraction pattern from Type 2 M-Wire contained additional peaks for martensite and R-phase, along with peaks for austenite. However, when M-Wire was examined by transmission electron microscopy, a heavily deformed martensitic structure was found [58]. The explanation is that the DSC peaks only reveal NiTi phases that are capable of undergoing transformation and that (stable) heavily deformed martensitic NiTi only produces weak x-ray diffraction peaks. Rotary instruments fabricated from M-Wire have been found to have similar A_f values, microstructures and Vickers hardness, so the machining process and other proprietary fabrication steps do not appear to markedly alter the inherent structure and properties of the starting blanks [60].

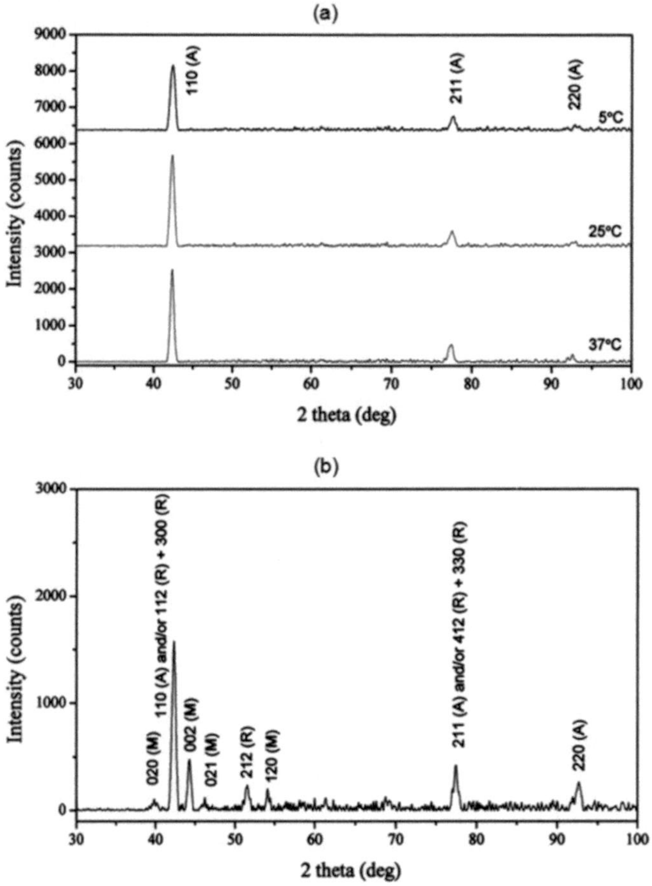

Figure 8. Micro-X-ray diffraction patterns for (a) Type 1 M-Wire and (b) Type 2 M-Wire. Peaks for austenite (A), martensite (M) and R-phase (R) are labeled. Reproduced from [58] with permission.

Recently, new nickel-titanium rotary instruments have been introduced, in which the wire blank is heated to an appropriate temperature for transformation to the R-phase and twisted, along with repeated heat treatment and other subsequent thermal processing; instruments have been characterized by conventional DSC and cantilever bending tests [61]. Another recent study has characterized several new nickel-titanium rotary instruments by DSC and conventional x-ray diffraction, along with optical and scanning electron microscopic examination of their microstructures, including use of energy-dispersive x-ray spectroscopic analyses (SEM/EDS), to investigate the martensitic microstructures and composition of precipitates [62]. Because of the potentially great commercial importance, development of new rotary instruments with improved clinical performance is expected to remain an area of intensive research, along with study of the role of heat treatment [63].

It is essential to appreciate the complexity of the physical metallurgy of the nickel-titanium alloys and the effects of the severe thermomechanical processing of the starting wire blanks, along with heat treatments and machining of the wire blanks, on the metallurgical structure. Transmission electron microscopy and electron diffraction remain the best techniques to gain insight into the instrument microstructures and elucidate the relationships with mechanical properties and clinical performance.

Author details

William A. Brantley
Division of Restorative, Prosthetic and Primary Care Dentistry,
Graduate Program in Dental Materials Science, College of Dentistry,
The Ohio State University, Columbus, OH, USA

Satish B. Alapati
Department of Endodontics, College of Dentistry, University of Illinois at Chicago, Chicago, IL, USA

5. References

[1] Anusavice KJ (editor). Phillips' Science of Dental Materials, 11th edition. St. Louis: Saunders/Elsevier Science; 2003.
[2] Powers JM, Sakaguchi RL (editors). Craig's Restorative Dental Materials, 12th edition. St. Louis: Mosby/Elsevier; 2006.
[3] Boyer DB, Edie JW. Composition of clinically aged amalgam restorations. Dent Mater 1990;6(3): 146-150.
[4] International Organization for Standardization: ISO 22674. Dentistry – Metallic materials for fixed and removable appliances; 2006.
[5] Watanabe I, Atsuta M, Yasuda K, Hisatsune K. Dimensional changes related to ordering in an AuCu-3wt%Ga alloy at intraoral temperature. Dent Mater 1994;10(6): 369-374.
[6] American National Standard/American Dental Association Specification No. 38. Metal-ceramic dental restorative systems: 2000 (Reaffirmed 2010). This specification is a modified adoption of ISO 9693:1999.
[7] American Dental Association, Council on Scientific Affairs. Revised classification system for alloys for fixed prosthodontics. [http://www.ada.org/2190.aspx]
[8] Rosenstiel SF, Land MF, Fujimoto J. Contemporary Fixed Prosthodontics, 4th edition. Mosby/Elsevier; 2006.
[9] Mackert JR Jr, Ringle RD, Fairhurst CW. High-temperature behavior of a Pd-Ag Alloy for porcelain. J Dent Res 1983;62(12): 1229-1235.
[10] Ohno H, Kanzawa Y. Structural changes in the oxidation zones of gold alloys for porcelain bonding containing small amounts of Fe and Sn. J Dent Res 1985;64(1): 67-73.
[11] Papazoglou E, Brantley WA, Mitchell JC, Cai Z, Carr AB. New high-palladium alloys: studies of the interface with porcelain. Int J Prosthodont 1996;9(4): 315-322.

[12] Papazoglou E, Brantley WA, Carr AB, Johnston WM. Porcelain adherence to high-palladium alloys. J Prosthet Dent 1993;70(5): 386-394.
[13] Vermilyea SG, Huget EF, Vilca JM. Observations on gold-palladium-silver and gold-palladium alloys. J Prosthet Dent 1980;44(3): 294-299.
[14] Carr AB, Cai Z, Brantley WA, Mitchell JC. New high-palladium casting alloys. Part 2. Effects of heat treatment and burnout temperature. Int J Prosthodont 1993;6(3): 233-241.
[15] Vermilyea SG, Cai Z, Brantley WA, Mitchell JC. Metallurgical structure and microhardness of four new palladium-based alloys. J Prosthodont 1996;5(4): 288-294.
[16] Guo WH, Brantley WA, Li D, Clark WA, Monaghan P, Heshmati RH. Annealing study palladium-silver dental alloys: Vickers hardness measurements and SEM microstructural observations. J Mater Sci Mater Med 2007;18(1): 111-118.
[17] Baran GR. The metallurgy of Ni-Cr alloys for fixed prosthodontics. J Prosthet Dent 1983;50(5): 639-650.
[18] Morris, HF, Asgar K, Rowe AP, Nasjleti CE. The influence of heat treatments on several types of base-metal removable partial denture alloys. J Prosthet Dent 1979;41(4): 388-395.
[19] Bridgeport DA, Brantley WA, Herman PF. Cobalt-chromium and nickel-chromium alloys for removable prosthodontics, Part 1: Mechanical properties. J Prosthodont 1993;2(3): 144-150.
[20] Waldmeier MD, Grasso, JE, Norburg, GJ, Nowak MD. Bend testing of wrought wire removable partial denture alloys. J Prosthet Dent 1996;76(5): 559-565.
[21] Donachie MJ Jr. Titanium: A Technical Guide, 2nd edition. Materials Park, OH: ASM International; 2000.
[22] Jeong Y-H, Choe H-C, Brantley WA. Nanostructured thin film formation on femtosecond laser-textured Ti–35Nb–xZr alloy for biomedical applications. Thin Solid Films 2010-2011;519(15): 4668-4675.
[23] Lee H, Dregia S, Akbar S, Alhoshan S. Growth of 1-D TiO_2 nanowires on Ti and Ti alloys by oxidation. J Nanomaterials 2010;502186 [DOI:10.1155/2010/503186].
[24] Brantley WA, Eliades T (editors). Orthodontic Materials: Scientific and Clinical Aspects. Stuttgart: Thieme; 2001.
[25] Verstrynge A, Van Humbeeck J, Willems G. In-vitro evaluation of the material characteristics of stainless steel and beta-titanium orthodontic wires. Am J Orthod Dentofacial Orthop 2006;130(4): 460-470.
[26] Goldberg AJ, Vanderby R Jr, Burstone CJ. Reduction in the modulus of elasticity in orthodontic wires. J Dent Res 1977;56(10): 1227-1231.
[27] Asgharnia MK, Brantley WA. Comparison of bending and tension tests for orthodontic wires. Am J Orthod 1986;89(3): 228-236.
[28] Khier SE, Brantley WA, Fournelle RA. Structure and mechanical properties of as-received and heat-treated stainless steel orthodontic wires. Am J Orthod Dentofacial Orthop 1991;99: 310-318.
[29] Backofen WA, Gales GF. Heat treating stainless steel wire for orthodontics. Am J Orthod 1951;21(2): 117-124.
[30] Funk AC. The heat treatment of stainless steel. Angle Orthod 1951;21(3); 129-138.
[31] Howe GL, Greener EH, Crimmins DS. Mechanical properties and stress relief of stainless steel orthodontic wire. Angle Orthod 1968;38(3): 244-249.

[32] Fillmore GM, Tomlinson JL. Heat treatment of cobalt-chromium alloys of various tempers. Angle Orthod 1979;49(2): 126-130.

[33] Goldberg J, Burstone CJ. An evaluation of beta titanium alloys for use in orthodontic appliances. J Dent Res 1979;58(2): 593-599.

[34] Burstone CJ, Goldberg AJ. Beta titanium: A new orthodontic alloy. Am J Orthod 1980;77(2): 121-132.

[35] Andreasen GF, Hilleman TB. An evaluation of 55 cobalt substituted nitinol wire for use in orthodontics. J Am Dent Assoc 1971;82(6): 1373-1375.

[36] Andreasen GF, Brady PR. A use hypothesis for 55 nitinol wire for orthodontics. Angle Orthod 1972;42(2): 172-177.

[37] Andreasen GF, Morrow RE. Laboratory and clinical analyses of nitinol wire. Am J Orthod 1978;73(2): 142-151.

[38] Burstone CJ, Qin B, Morton JY. Chinese NiTi wire — a new orthodontic alloy. Am J Orthod 1985;87(6): 445-452.

[39] Miura F, Mogi M, Ohura Y, Hamanaka H. The super-elastic property of the Japanese NiTi alloy wire for use in orthodontics. Am J Orthod Dentofacial Orthop 1986;90(1): 1-10.

[40] Khier SE, Brantley WA Fournelle RA. Bending properties of superelastic and nonsuperelastic nickel-titanium orthodontic wires. Am J Orthod Dentofacial Orthop 1991;99(4): 310-318.

[41] Fletcher ML, Miyake S, Brantley WA, Culbertson BM. DSC and bending studies of a new shape-memory orthodontic wire. J Dent Res 1992;71(Spec Iss A): 169, Abstract No. 505.

[42] Duerig TW, Melton KN, Stöckel D, Wayman CM (editors). Engineering Aspects of Shape Memory Alloys. London: Butterworth-Heinemann; 1990.

[43] Kusy RP. A review of contemporary archwires: Their properties and characteristics. Angle Orthod 1997;63 (3): 197-207.

[44] Miura F, Mogi M, Ohura Y. Japanese NiTi alloy wire: Use of the direct electric resistance heat treatment method. Eur J Orthod 1988;10(3): 187-191.

[45] Bradley TG, Brantley WA, Culbertson BM. Differential scanning calorimetry (DSC) analyses of superelastic and nonsuperelastic nickel-titanium orthodontic wires. Am J Orthod Dentofacial Orthop 1996;109(6): 589-597.

[46] Brantley WA, Iijima M, Grentzer TH. Temperature-modulated DSC provides new insight about nickel-titanium wire transformations. Am J Orthod Dentofacial Orthop 2003;124(4): 387-394.

[47] International Organization for Standardization: ISO 15841. Dentistry — Wires for use in orthodontics; 2006. ANSI/ADA Specification No. 32 — Orthodontic Wires: 2006 is an identical adoption of this ISO standard.

[48] Khier SE. Structural Characterization, Biomechanical Properties, and Potentiodynamic Polarization Behavior of Nickel-Titanium Orthodontic Wire Alloys [Ph.D. Dissertation]. Milwaukee, WI, USA: Marquette University; 1988.

[49] Brantley WA, Guo W, Clark WA, Iijima M. Microstructural studies of 35°C copper Ni-Ti orthodontic wire and TEM confirmation of low-temperature martensite transformation. Dent Mater 2008;24(2): 204-210.

[50] Canalda-Sahli C, Brau-Aguadé E, Sentís-Vilalta J. The effect of sterilization on bending and torsional properties of K-files manufactured with different metallic alloys. Int Endod J 1998;31(1): 48-52.
[51] Walia H, Brantley WA, Gerstein H. An initial investigation of the bending and torsional properties of Nitinol root canal files. J Endod 1988;14: 346-351.
[52] Thompson SA. An overview of nickel-titanium alloys used in dentistry. Int Endod J 2000;33(4): 297-310.
[53] Alapati SB, Brantley WA, Svec TA, Powers JM, Nusstein JM, Daehn GS. SEM observations of nickel-titanium rotary endodontic instruments that fractured during clinical Use. J Endod 2005;31(1): 40-43.
[54] Parashos P, Messer HH. Rotary NiTi instrument fracture and its consequences. J Endod 2006;32(11): 1031-1043.
[55] Alapati SB, Brantley WA, Iijima M, Schricker SR, Nusstein JM, Li U-M, Svec TA. Micro-XRD and temperature-modulated DSC investigation of nickel-titanium rotary endodontic instruments. Dent Mater 2009;25(10): 1221-1229.
[56] Tripi TR, Bonaccorso A, Rapisarda E, Tripi V, Condorelli GG, Marino R, Fragalà I. Depositions of nitrogen on NiTi instruments. J Endod 2002;28(7): 497-500.
[57] Zinelis S, Darabara M, Takase T, Ogane K, Papadimitriou GD. The effect of thermal treatment on the resistance of nickel–titanium rotary files in cyclic fatigue. Oral Surg Oral Med Oral Pathol Oral Radiol Endod 2007;103(6): 843-847.
[58] Alapati SB, Brantley WA, Iijima M, Clark WA, Kovarik L, Buie C, Liu J, Johnson WB. Metallurgical characterization of a new nickel-titanium wire for rotary endodontic instruments. J Endod 2009;35(11): 1589-1593.
[59] Brantley WA, Svec TA, Iijima M, Powers JM, Grentzer TH. Differential scanning calorimetric studies of nickel titanium rotary endodontic instruments. J Endod 2002;28(8): 567-572.
[60] Liu J. Characterization of New Rotary Endodontic Instruments Fabricated from Special Thermomechanically Processed NiTi Wire [Ph.D. Dissertation]. Columbus, OH, USA: Ohio State University; 2009.
[61] Hou X, Yahata Y, Hayashi Y, Ebihara A, Hanawa T, Suda H. Phase transformation behaviour and bending property of twisted nickel-titanium endodontic instruments. Int Endod J 2011;44(3): 253-258.
[62] Shen Y, Zhou HM, Zheng YF, Campbell L, Peng B, Haapasalo M. Metallurgical characterization of controlled memory wire nickel-titanium rotary instruments. J Endod 2011;37(11): 1566-1571.
[63] Yahata Y, Yoneyama T, Hayashi Y, Ebihara A, Doi H, Hanawa T, Suda H. Effect of heat treatment on transformation temperatures and bending properties of nickel-titanium endodontic instruments. Int Endod J 2009;42(7): 621-626.
[64] Otsuka K, Ren X. Physical metallurgy of Ti–Ni-based shape memory alloys. Prog Mater Sci 2005;50(5): 511-678.

Chapter 2

Microstructure-Property Relationship in Advanced Ni-Based Superalloys

Hiroto Kitaguchi

Additional information is available at the end of the chapter

http://dx.doi.org/10.5772/52011

1. Introduction

Ni based superalloys have been developed more or less empirically over the past 60 years from a simple Ni-Cr matrix to the present multi element and phase systems[1], having a fully austenitic face centred cubic (fcc) structure which maintains a superior tensile, fatigue and creep properties at high temperature to a body centred cubic (bcc) alloy[1]. One of the major applications of Ni superalloys is gas turbine engines. They comprise over 50% of the weight of advanced aircraft engines and include wrought and cast turbine blades and powder metallurgy (P/M) route turbine discs [1].

One of the most important goals of engine design is increasing turbine entry temperature (TET): the temperature of the hot gases entering the turbine arrangement [2]. This implies that the resistance against the environmental attack, i.e. high temperature, under a severe mechanical force is the priority challenge and indeed Ni based superalloys are used in the hottest as well as the highest tensile pressure of the gas turbine engine component as shown in the schematic diagram in Fig. 1. Nowadays, for the advanced cast single crystal superalloys in the turbine blades, the alloy capability exceeds 1,000ºC [2]. In this chapter, the polycrystalline Ni superalloys, which have slightly less temperature capability up to 800°C, applied in the turbine discs and the adjoined shafts, will be introduced focusing on their microstructures correlating with the mechanical properties.

2. Microstructure (second phases)

2.1. Hardening precipitates

From the point of view of microstructure, Ni superalloys are complex [4]. The fcc matrix, known as γ, mainly consists of nickel, cobalt, iron, chromium and molybdenum. The strength of superalloys are conferred by the hardening precipitates known as γ' (Ni_3Al based $L1_2$

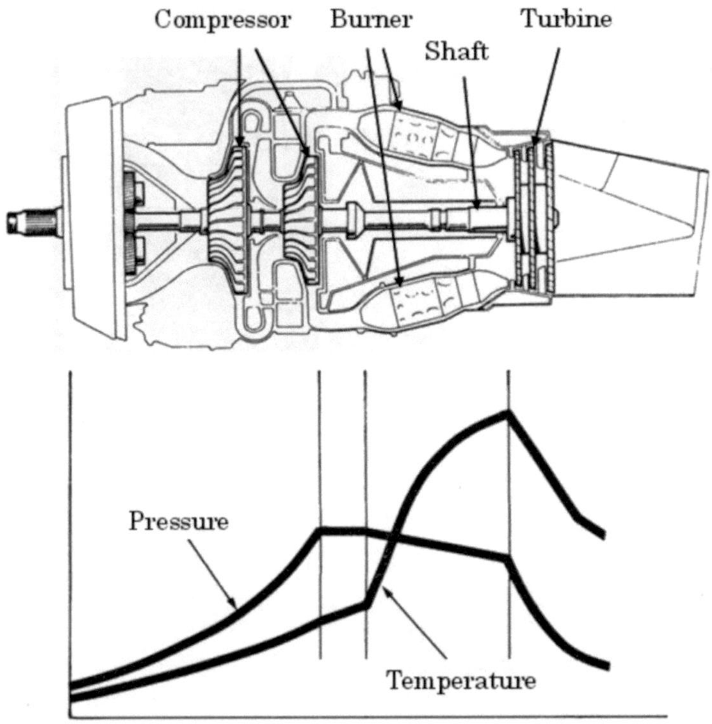

Figure 1. (a) Schematic diagram of a turbine engine Ref. [3]. (b) Schematic diagram of the temperature and pressure gradients throughout the engine component correlating with the diagram (a) Ref. [4]

structure) (Fig. 2). In some nickel – iron superalloys such as IN718 and IN706, which contain niobium, they are hardened by γ'' (Ni_3Nb based $D0_{22}$ structure) (Fig. 3) [2]. Homogeneously distributed coherent hardening precipitates confer excellent tensile and fatigue life properties at high temperatures. Their volume fraction is controlled by the nominal chemical composition. The size and the morphology are controlled by the process and their crystallographic relations with γ matrix. The precipitates arise close to the solvus temperature grow larger which subsequently restrict the grain growth pinning grain boundaries (Fig. 4). On the other hand, the precipitates arise at lower temperature such as during cooling after heat treatment stay small (Fig. 4 (left hand side of the image)). γ' has the perfect coherency with the γ matrix, hence their morphologies are mostly sphere, whereas γ'' has a tall crystal unit tetragonal structure where *a* axis has the identical lattice parameter with the γ matrix but *c* axis has nearly double the length of the γ, hence γ'' always precipitate with the perfect coherency on the basal plane with γ and grow along the longitudinal direction (Fig. 5).

2.2. Carbides and borides

Carbon and boron are added as a grain boundary strengthener by segregating in the grain boundaries and forming carbides and borides. They are believed to be formed during solidification, aging treatment which strengthen grain boundaries at elevated temperatures but the ones arising during service must be controlled carefully since they can impair properties [4].

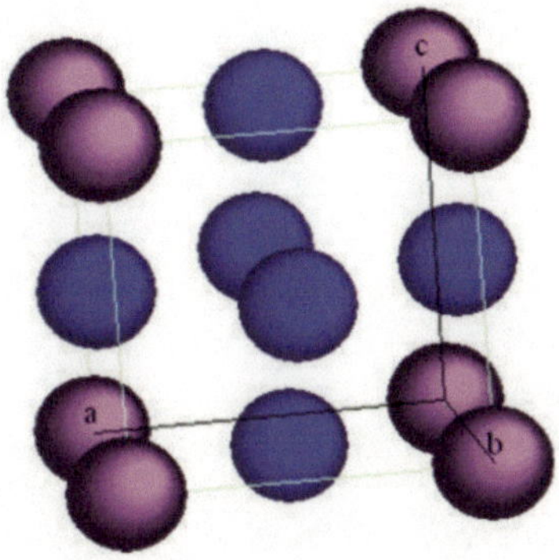

Figure 2. γ' $L1_2$ structure. Ni atoms are blue and Al purple

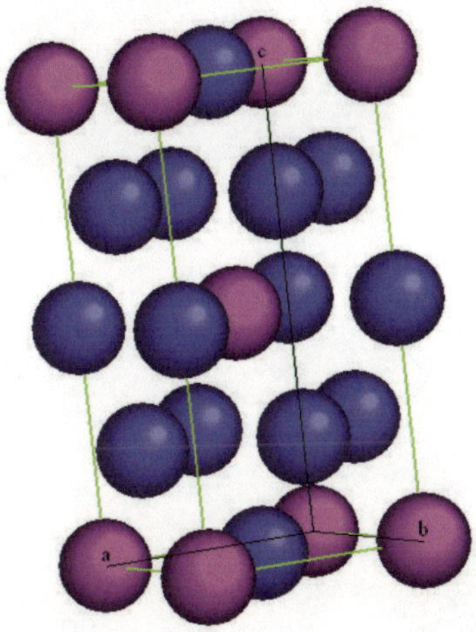

Figure 3. γ'' $D0_{22}$ structure. Ni atoms are blue and Nb, Al and Ti purple

Carbides are traditionally classified by their chemical composition, mainly MC, M_6C and $M_{23}C_6$, where M stands for metal elements such as Ti, Cr, Nb, Mo, Hf and Ta [4].

MC carbides are usually coarse (Fig. 6), having a fcc densely packed structure [4]. Ti, Nb, Hf and Ta are the main metal elements. They are very strong and are normally considered to be some of the most stable compounds in nature, justified by their high precipitation and melting temperature: they are believed to precipitate during processing shortly after solidification of the superalloy [4]. They usually have little or no orientation relationship with the alloy matrix [4].

M_6C carbides have a complex cubic structure and they precipitate when the alloy contains highly refractory elements, for example Mo and W. These carbides are believed to be the product of MC carbide decomposition during service or relatively high heat treatment between 815 and 980ºC [4]. The examples of the micrographs of M_6C can be found in Ref. [5, 6].

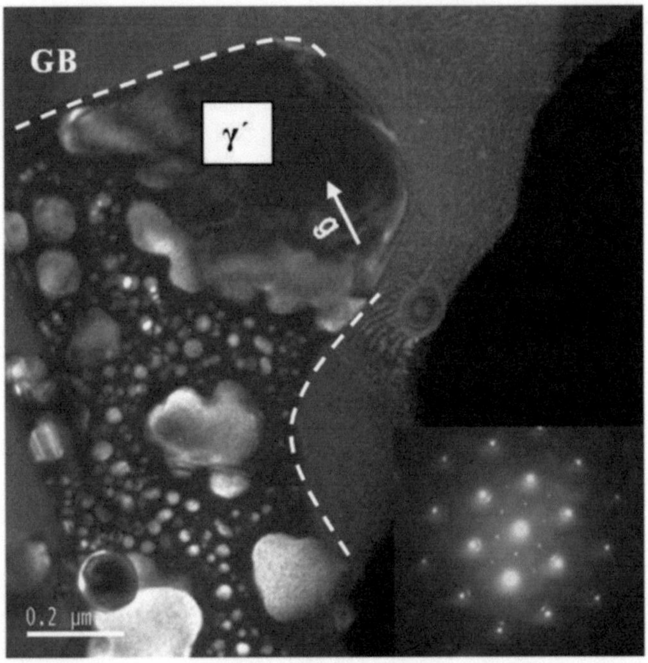

Figure 4. TEM dark field (DF) image. γ' pinning grain boundary, shown by the white dashed line. The small spherically shaped precipitates inside the grain are also γ'. ($g = 01\bar{1}$ B = [111])

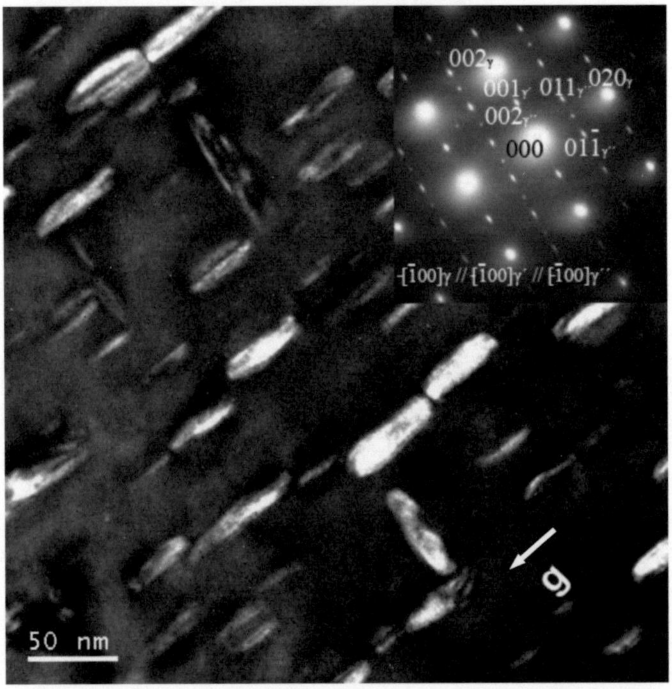

Figure 5. TEM DF image of the γ'' in IN718. The growth direction is c axis parallel to the a axis of γ (($g_{\gamma''} = 0\ 0\ 2$) B = [100]).

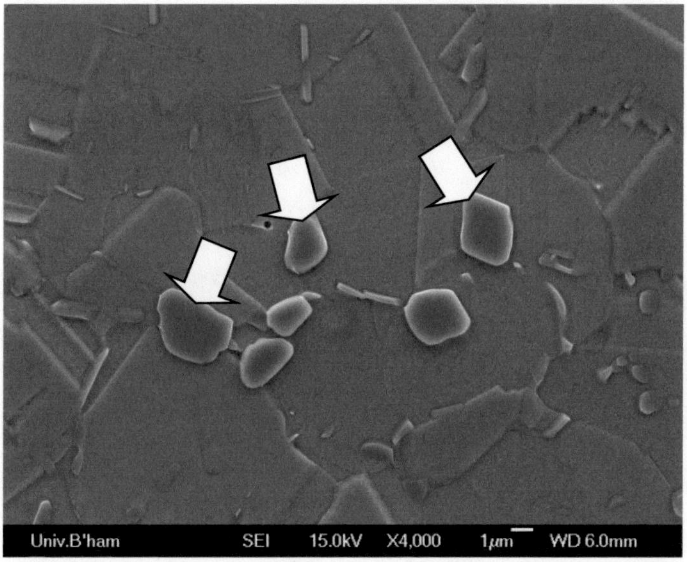

Figure 6. Coarse Nb and Ti based carbide in IN718

$M_{23}C_6$ carbides (Fig. 7) form mainly along grain boundaries at a relatively low temperature for carbides: between 760 and 980ºC. The crystal structure is complex cubic structure. The lattice parameter is exactly three times larger than γ matrix, hence they precipitate with cube-cube orientation with the matrix (Fig. 8). They are believed to form either by the decomposition of MC or M_6C or they nucleate directly on the grain boundaries. They are known as having a high content of Cr. $M_{23}C_6$ carbides have a significant effect on Ni based superalloy properties [4] since they are profuse in alloys with moderate to high Cr content [4]

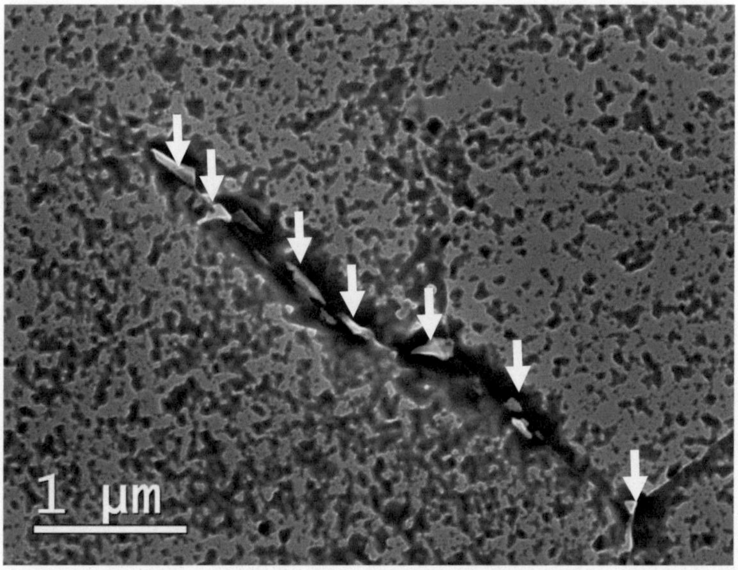

Figure 7. Fine $M_{23}C_6$ type carbides precipitate along the grain boundary running diagonally.

and are controversial carbides. Firstly, this is because their different morphologies: the blocky shaped ones at grain boundaries have a beneficial effect on rupture strength; on the contrary the film ones are regarded as promoting early rupture failure [4]. Secondly, this is because that they make a Cr depleted zone (Fig. 9) around the precipitate. In this area, it is difficult to form a protective oxide, namely Cr_2O_3, due to lack of Cr.

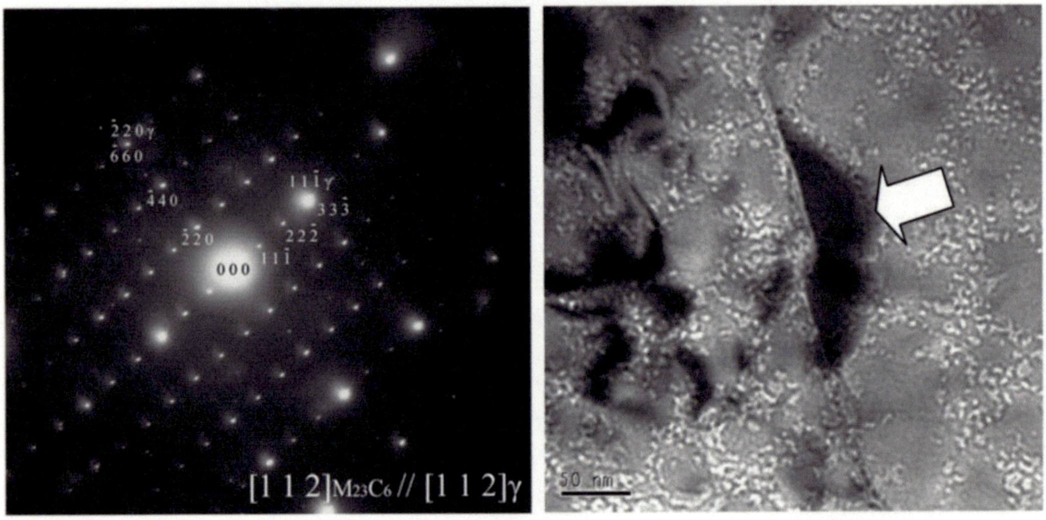

Figure 8. $M_{23}C_6$ and γ matrix perfect coherent diffraction pattern (left) and the bright field image from another beam direction to make $M_{23}C_6$ outstanding (right)

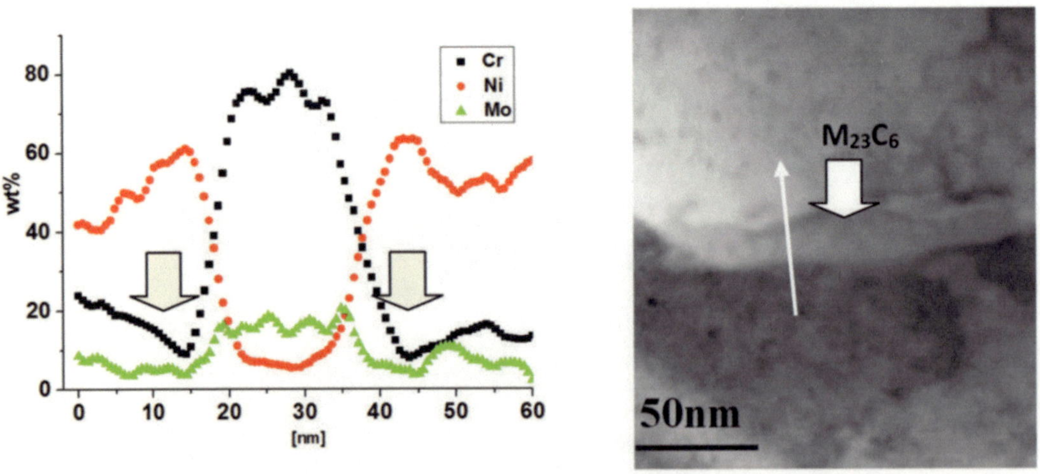

Figure 9. Left: STEM EDX line scan results across $M_{23}C_6$ revealed the Cr depletion as indicated by the arrows (Cr nominal composition is 15 wt%). Right: STEM bright field image of the $M_{23}C_6$ (the thin arrow shows the length and the direction of the EDX line scan)

It is broadly acknowledged that boron segregation along the grain boundary increases the cohesive strength of the grain boundaries. The role of borides is, however, still under open discussion. Those so far identified have a base centred tetragonal (BCT), M_3B_2 [4] or M_5B_3 [7] formula, where M is typically a refractory element, namely Mo or Cr. They appear as various shapes such as blocky to half-moon [4]. The examples shown in Fig. 10 were found in an advanced polycrystalline Ni superalloy after a thermal exposure at 980°C for 1 hour.

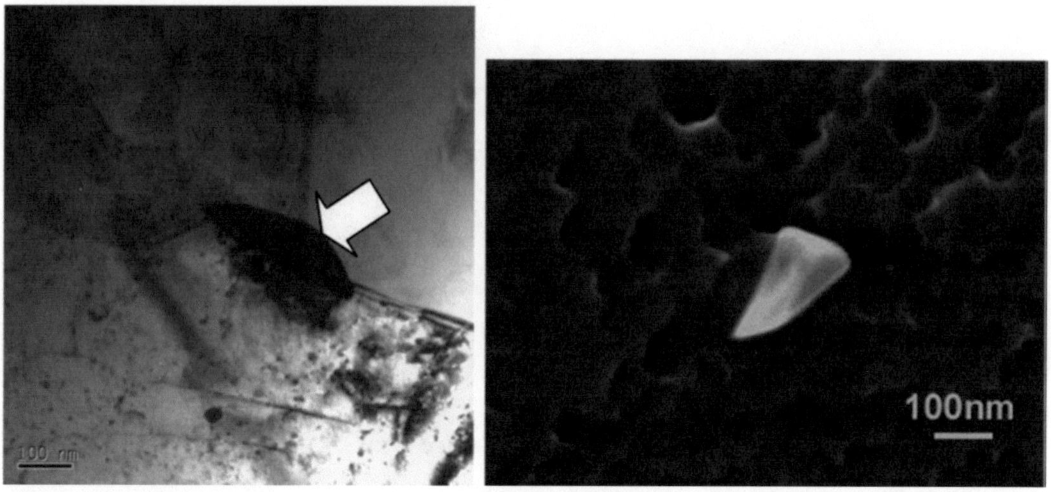

Figure 10. Some examples of M_5B_3 type boride appeared in TEM bright field (left) and in SEM (right)

2.3. Other phases

Adding excess quantity of refractory elements, such as Mo, W and Re, promotes the precipitation of hard intermetallic phases [2], so called TCP phase, which are believed to deteriorate the alloy ductility [4] and the creep life [8]. In the ternary phase diagrams for superalloy elements, such as Ni-Cr-Mo, there are two phase spaces: one is austenite (γ) fcc and the other is bcc [4]. Between these two major fields, a band of numerous small phase volumes can be identified such as σ, μ, R and so on [4], which are characterized firstly as having a high and uniform packing density of atoms[2] and secondly as having complex crystal structures [2], either hcp, body centred tetragonal or rhombohedral. With the careful control of these refractory elements, TCP phases occur after a long time service or a prolonged heat treatment [9]. Some are believed to be the products of transformation from another beneficial phase: for example $\eta(Ni_3X)$ results from γ' [4] and σ has the same crystal structure as that of $M_{23}C_6$, but without the carbon atoms. The example of σ phase shown in Fig. 11 was found to be Cr, Mo and Co based chemistry after a thermal exposure at 720°C for 1,100 hours in a newly developed advanced Ni superalloy. The second phases introduced above and some other important second phases for the Ni superalloy microstructure are summarized in Table 1.

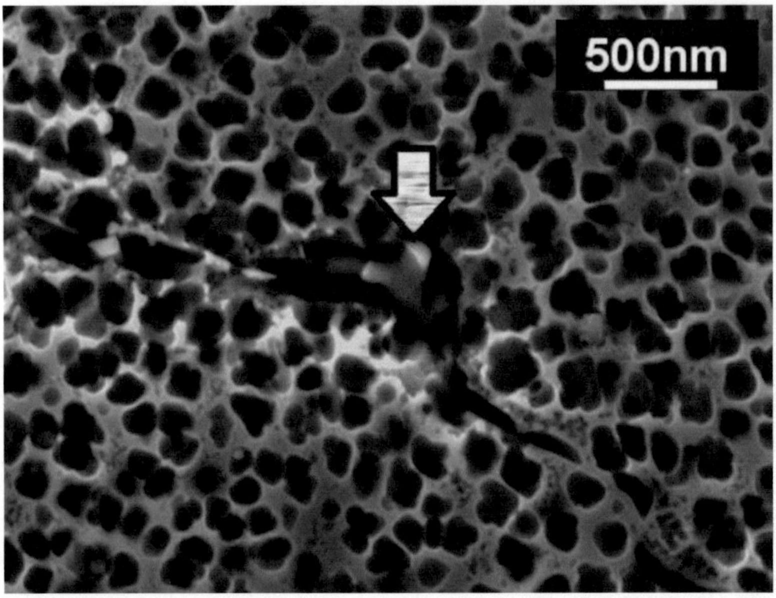

Figure 11. Sigma (σ) phase precipitates on the grain boundary running diagonally from top left to bottom right

Phase	Prototype	Pearson symbol	Strukturbericht symbol	Lattice [nm]	Chemical Composition (Appx)
γ'	Cu_3Au	cP4	$L1_2$	a 0.36	$(Ni\ Co)_3(Al\ Ti)$
γ''	Al_3Ti	tI8	$D0_{22}$	a 0.36 c 0.74	$(Ni\ Fe)_3(Nb\ Al\ Ti)$
MC	NaCl	cF8	B1	a 0.44	(Ti Ta)C or TiC, TaC, NbC, WC
M_6C	Fe_3W_3C	cF112	$E9_3$	a 1.11	$(Mo\ Cr\ W)_6C$
M_7C_3	Cr_7C_3	oP40	$D10_1$	a 0.45 b 0.70 c 1.21	Cr_7C_3
$M_{23}C_6$	$Cr_{23}C_6$	cF116	$D8_4$	a 1.07	$Cr_{21}Mo_2C_6$
M_5B_3	Cr_5B_3	tI32	$D8_1$	a 0.55 c 1.06	$(Cr\ Mo)_5B_3$
M_3B_2	Si_2U_3	tP10	$D5_a$	a 0.60 c 0.32	$(Mo\ Cr)_3B_2$
σ	CrNi	tP30	$D8_b$	a 0.88 c 0.46	Cr Mo Co based
δ	Cu_3Ti (β)	oP8	$D0_8$	a 0.51 b 0.43 c 0.46	Ni_3Nb
η	Ni_3Ti	hP16	$D0_{24}$	a 0.51 c 0.83	$Ni_3(Ti\ Ta)$
μ	Fe_7W_6	hR13	$D8_5$	a 0.48 c 2.5	Mo Co based

Table 1. Summary of second phases in the polycrystalline Ni based superalloys [10] The lattice parameter may vary (less than 5%) by changing chemical composition.

3. Microstructures and mechanical properties

It is worth noting the microstructure related mechanical properties in detail. We will discuss briefly how microstructure affects various mechanical properties in polycrystalline Ni superalloys.

Altering grain sizes results in various effects with regard to the different mechanical properties. Tensile and fatigue life properties are optimized by a fine grain microstructure, on the other hand, good creep and fatigue crack growth properties at elevated temperature are favoured by a coarse grain microstructure [2]. The former is a result of grain orientation and stress concentration by dislocation movement along the slip plane [2]. The latter is about intergranular crack propagation susceptibility. For example, Bain *et al* [11] showed the significance of the grain size for the crack growth rate using UDIMET720. Testing at 650°C, the crack growth rate reduced by more than two orders of magnitude by changing the size from 20 to 350 μm in diameter. (Fig. 12).

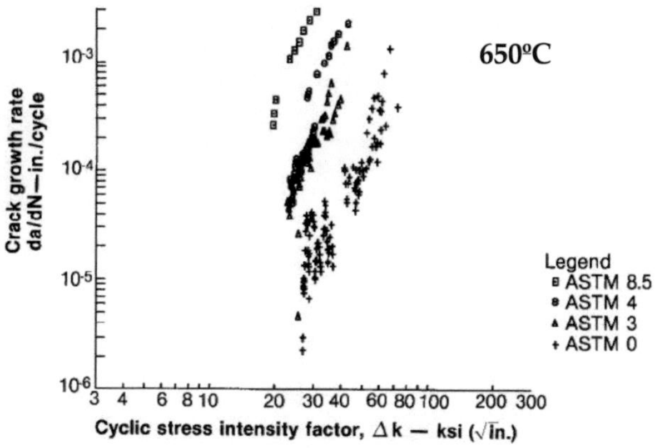

Figure 12. UDIMET 720 fatigue crack growth rate for different grain sizes (ASTM grain size between 0 and 8.5: 360μm and 19μm in diameter) tested at 650ºC [11]

The size of the hardening precipitates significantly affects the yield strength of the material via the interaction between the precipitate and the dislocation. If the precipitates are large, dislocation bowing around the precipitates becomes dominant; for small sized precipitates, dislocation cutting becomes dominant.

For bowing

$$\tau = \frac{G * b}{L - 2r} \tag{1}$$

and for cutting

$$\tau = \frac{r * \gamma * \pi}{b * L} \tag{2}$$

τ is the strength of the material, G is the shear modulus, b is the magnitude of the Burgers vector, L is the distance between the hardening precipitates, r is the radius of the precipitates and γ is the surface energy. In general in Ni-Al binary system, the optimum size to maximize the strength is found to be around 5 - 30 nm in diameter (Fig. 13). The size of the precipitates also affects the creep strain as shown in Fig. 14. In their study [12], the size of the precipitate was changed by changing the heat treatment temperature and time and found that the smaller the precipitate the slower the creep strain rate is, which is achieved via the smaller γ' - γ' channel width [12, 13].

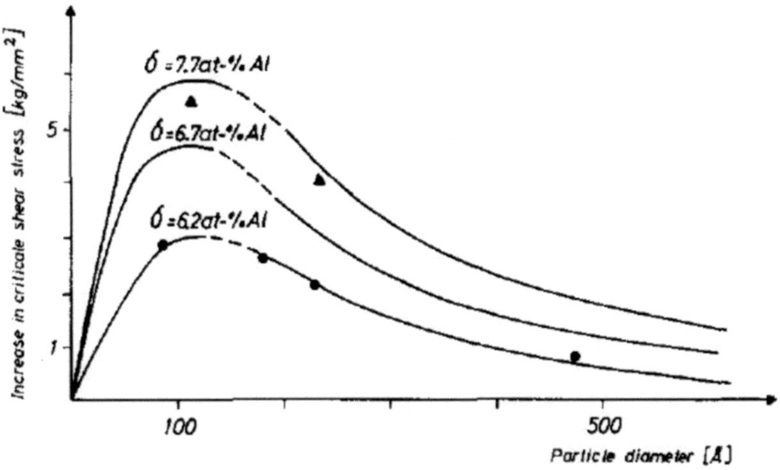

Figure 13. γ' particle diameter against the critical shear stress in Ni-Al system [28]

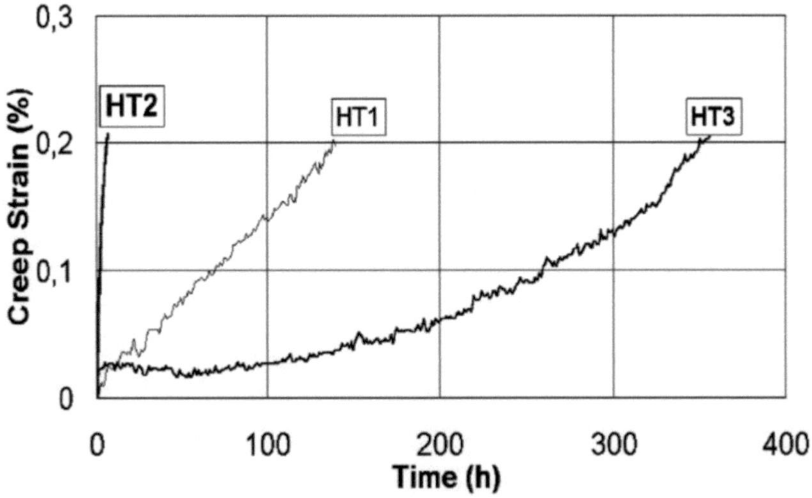

Figure 14. Creep strain tested at 700ºC for different heat treatments (HT1, HT2 and HT3) The size of γ': HT2>HT1>HT3 [12].

It is well known fact that in general both the yield strength and the creep rupture strength increases by increasing the hardening precipitate volume fraction [2]. Historically, low cycle fatigue life was the main concern for turbine disc alloys, but fatigue crack growth rate and damage tolerant design have attracted more attention over the last two decades [11, 14].

They can be strongly influenced not only by the size of the grains as introduced above, but also by the size of the precipitates; the striking results were shown in Ref. [15, 16]. The results show that the larger the hardening precipitates the better the crack growth property. However, this conflicts with the creep life property as mentioned above. Research on damage tolerant design originally started to investigate the grain boundary chemistry since fast crack growth (FCG) is always observed with intergranular cracks and tends to disappear at low temperature. Additionally, transgranular ductile cracking replaces intergranular crack when the tests carried out in the reduced oxygen partial pressure [17, 18] (Fig. 15) Thus, FCG embrittlement has been attributed to oxidation [11, 19]. Grain boundary engineering has been explored by changing the morphology of the grain boundary. For example, Ref. [15, 20] reported a complex grain boundary geometry, so called 'serrated' (Fig. 16), by slow cooling after solution treatment. The result showed slower intergranular crack growth rate than with a normal grain boundary [15]. However, the improvement above did not account for the property change by the different size of the hardening precipitate mentioned above. The fast intergranular crack growth at high temperature in superalloys added a new dimension after intensive studies with regard to the correlation between the hardening precipitate distribution and the crack growth rate. Ref. [15, 16, 21] claimed that the prevention of stress relaxation of the crack tip by the hardening precipitates can increase the crack growth rate. Some experimental work support the idea, for example Andieu *et al* [22] carried out a unique dwell fatigue crack propagation test where oxygen was introduced in different phases of the low cycle fatigue crack growth test and found that it is potent for the fast crack growth when oxygen is introduced at the beginning of the loading rather than introducing in the later part of the loading. This may imply that the oxidation at the crack tip happens during the stress concentrated at the crack tip. Molins et al [23, 24] concluded that the local microstructure at the crack tip, which can be controlled by an appropriate heat treatment against the stress accumulation, can significantly affect the crack propagation behaviour in Ni superalloys. This conclusion recalls an arguable grain boundary microstructure feature, namely the precipitate free zone (PFZ). One suggested that the PFZ would promote plastic deformation and fracture [25, 26]. Another suggested that the PFZ in some nickel alloys is beneficial for crack tip stress relaxation [27].

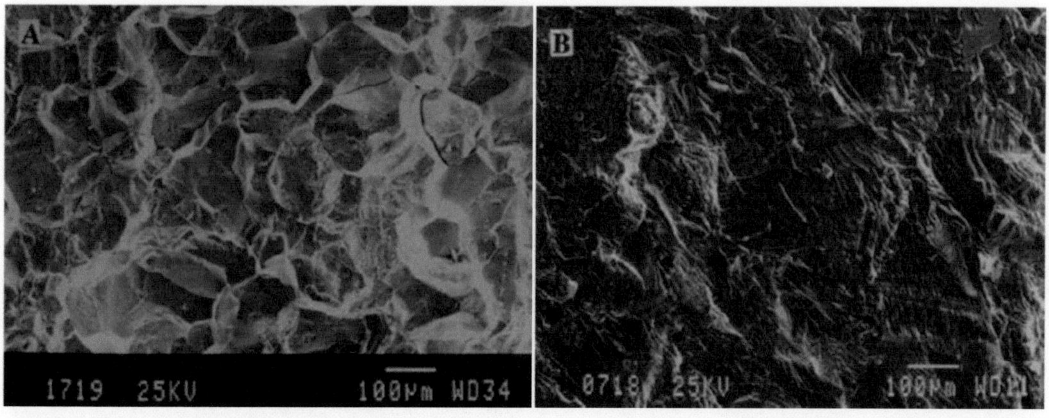

Figure 15. Typical intergranular (left) and transgranular (right) fracture surfaces. Alloy 718 tested at 650ºC in air (left) and vacuum (right) [18]

These findings above suggest that not only the macroscopic structure such as the grain size and the distribution of the hardening precipitates, but also the microscopic structure, such as the grain boundary shape and the relationship with the hardening precipitates, can significantly affect the mechanical properties.

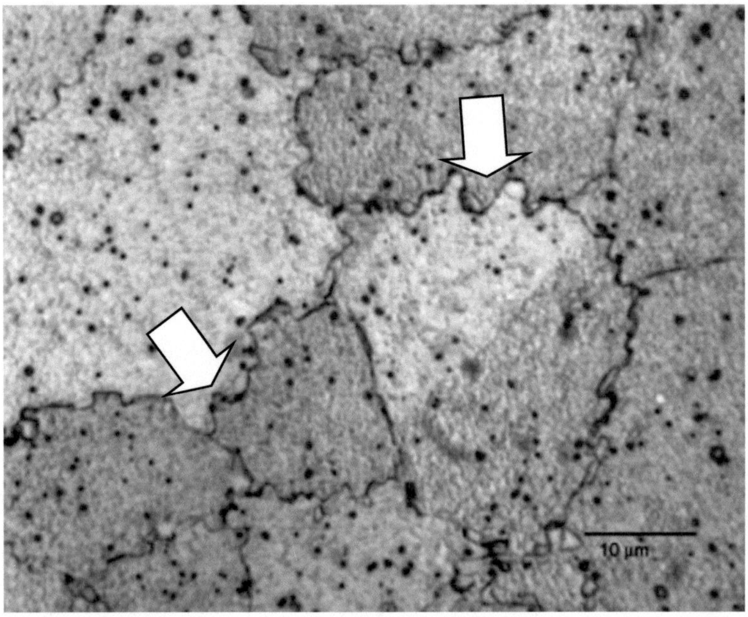

Figure 16. Optical microscopy image of serrated grain boundaries. The arrows indicate the serrated boundaries [15]

4. Polycrystalline superalloy grain boundary structure

The details of the Ni superalloy grain boundary microstructure will be demonstrated in this section. Particular attention will be paid to the relationship between the hardening precipitates and the high grain boundaries. Fig. 17 shows the STEM bright field image of the grain boundary and the hardening precipitate morphology in an advanced polycrystalline superalloy. The grain boundary running top left to bottom right cuts through γ'. This was confirmed by the conventional TEM image analysis combining with the crystallographic analysis that the either side of the γ' keeps the coherency with the matrix (Fig. 18). With respect to the morphologies of γ' on the grain boundaries, it is the same as those inside the grains. It has, however, two different crystallographic orientations keeping the coherency with the either side of the matrix. This morphology is believed to form during the process with the high boundary mobility [29]. There are at least four different possibilities of interactions between the migrating grain boundaries and the precipitates, which are illustrated in Fig. 19. Following Fig. 19,

a. the boundary migrates with no effect on the precipitates; the precipitates thus become incoherent after the migrating grain boundary passes through them.

b. the precipitates dissolve in contact with migrating boundary and reprecipitate coherently within the new grain.
c. the grain boundary is held by the coherent precipitates which then coarsen, leading to complete halting of the boundary movement.
d. the grain boundary can pass through the coherent precipitate which undergoes the same orientation change as the grain surrounding it and thereby retains the coherent low-energy interface between the precipitate and the matrix.

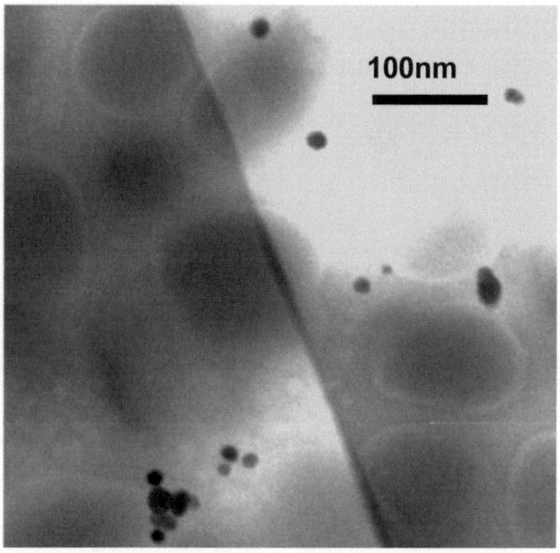

Figure 17. General aspect of the high angle grain boundary and γ' (dark spheres). The grain boundary is running diagonally from top left to bottom right

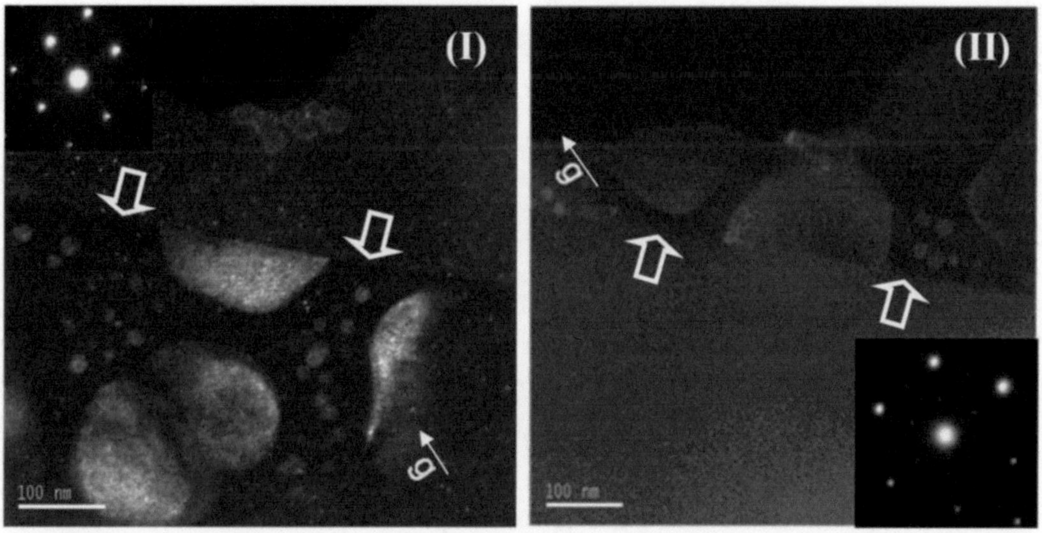

Figure 18. A crystallographic analysis of the cutting γ'. The dark field images of the lower and the upper grain are in (I) and (II), respectively. The white arrows indicate the grain boundary

With regard to theory a), this is often observed with high solvus temperature precipitates, such as carbides and oxides. b) is not applicable in this study. This can be, however, the case for less γ' volume fraction superalloys with small amount of nominal Al content such as Nimonic PE16 or the case in high temperature very close to the γ' solvus. With regard to c), it can be applicable in the case of the larger γ' such as the one in Fig. 4. d) is relevant to explain the results of Fig. 17 and Fig. 18. As the grain boundary impinges on the γ', the grain boundary apparently cuts off γ'. Firstly, the interface free energy between γ' and the grain boundary increases. This results in dissolving the γ' at the interface but due to the supersaturation of γ' formers, such as Al and Ti, γ' immediately re - nucleates in the next grain coherent with the next grain discontinuously [30, 31]. Thus, this phenomenon can be concluded the result of γ' dissolution and subsequent (discontinuous) precipitation.

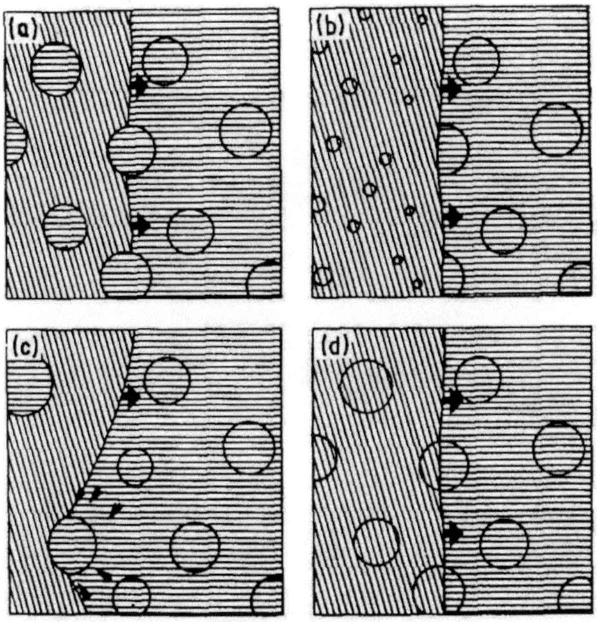

Figure 19. The possible interactions between the migrating grain boundary and the precipitate [29]

Another example is forming a precipitate free zone (PFZ) as shown in Fig. 20 in IN718. γ' and γ'' coexist in IN718, however, γ'' denude along the grain boundary and form a γ'' PFZ. On the contrary, the minor hardening precipitate in IN718: γ' exist along the grain boundary. Vacancy deficiency is one of the causes of the PFZ along grain boundaries due to lack of the nucleation sites as grain boundary acts as a good vacancy sink [32]. One of the important factors to create the PFZ in the superalloys can be the difference of the interfacial free energy, i.e. the free energy between γ'-matrix and γ''-matrix. The γ''-γ'' nucleus channel distance along the grain boundary can be larger than the critical distance to aggregate two γ'' nuclei. On the other hand, the critical distance for the γ' is smaller than that of γ'' or γ' can nucleate their precipitate independently as γ' has smaller interfacial energy. Thus, the γ' nuclei can grow and form precipitates along the grain boundary but not for γ'' and the γ'' PFZ arises.

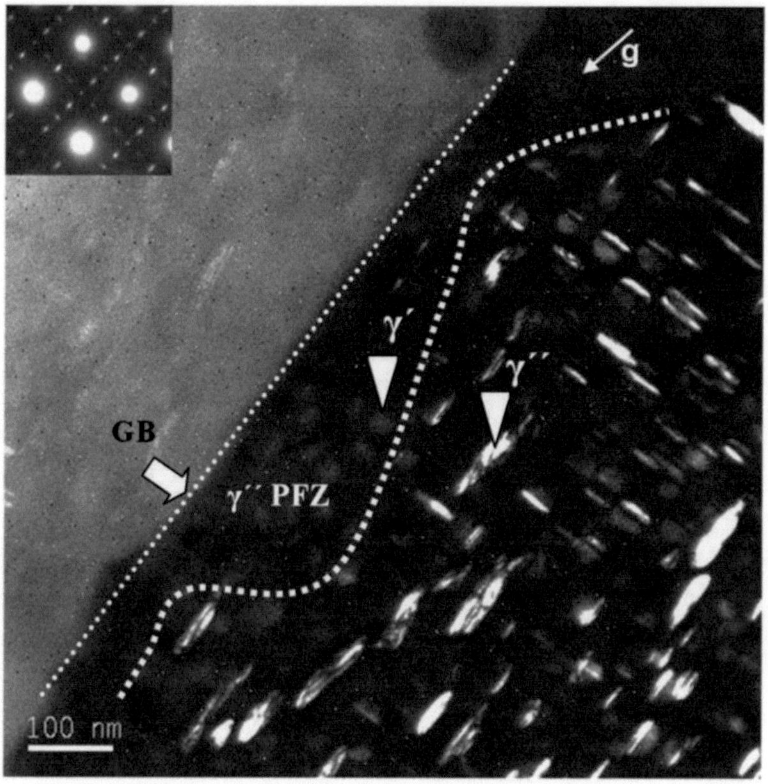

Figure 20. γ'' PFZ appeared in IN718, although γ' still exist in the γ''PFZ adjacent to the grain boundary

5. High temperature oxidation along grain boundaries

The context of 'High temperature' in this section is the temperature range of 600-700°C which is the high temperature regime of the disc in the turbine engine application.

It is more than a half century ago, the investigation of the oxidation assisted fast crack growth started. Cr is believed to be an important element for the oxidation assisted crack growth. For example, as shown in Fig. 21, crack growth tests were conducted under various oxygen partial pressure on Ni-Cr binary alloys with 5, 20 and 30 wt% of Cr. It was found that the higher the Cr content the higher the transition oxygen partial pressure from transgranular to intergranular cracking is. The highest Cr content alloy: Ni-30wt%Cr did not show a transition pressure. Oxidation process on a freshly exposed alloy surface had been characterised intensively and well understood. As illustrated in Fig. 22, both Ni and Cr oxide formation takes place at the beginning of the oxidation [33]. This is particularly important for alloys on the borderline between protective and non-protective behaviour [34]. But in the early stage, the fast kinetics Ni oxide grows quickly and dominates the oxide. In general, there are two types of oxidation: the cation diffusion type and the anion diffusion type [34] . The difference between the two is the movement of the ions; the former involves cation (metal ion) transport, the latter anion (oxygen ion) transport. For the cation diffusion type the oxides form between the oxide and the free surface, but, for the anion diffusion, the oxides form between the oxide and the metal interface. During the transient stage which corresponds to the middle of the illustrations in Fig. 22, Cr_2O_3 particles are embedded inside the NiO layer. As NiO grows and the oxidation rate becomes slower, Cr_2O_3 soon establishes its own layer underneath the NiO layer, which implies that Cr_2O_3 can be the anion diffusion. Eventually the Cr_2O_3 layer is completed, where the layer prevents further diffusion of oxygen into the alloy, called passivation [34].

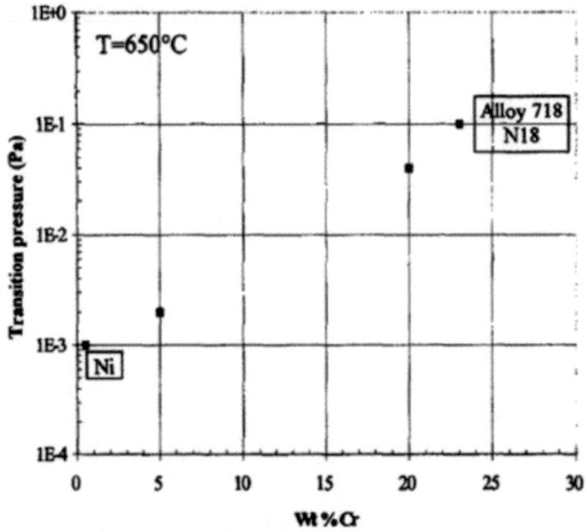

Figure 21. Transition oxygen partial pressure from the transgranular to intergranular cracking against the Cr concentration in Ni alloys after crack propagation tests at 650°C [24]

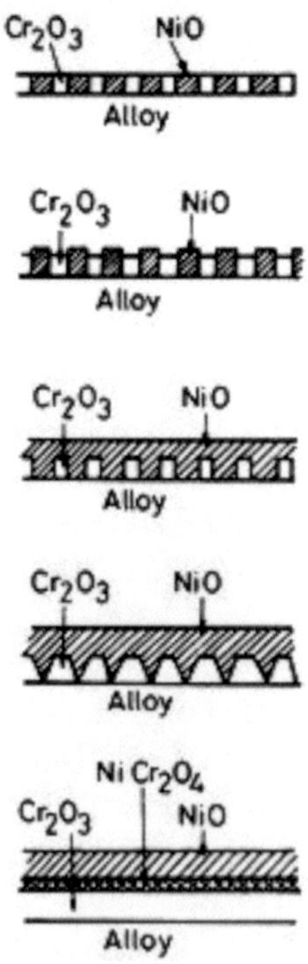

Figure 22. Schematic diagram of the oxidation process (from top to bottom) of the freshly exposed Ni alloy [33]

Nevertheless, the oxidation process at (ahead) of the crack tip has remained under debate. This is due to the technical difficulties of studying the microstructure of such small regions, which is predictable taking consideration of the size of the crack tip: less than a micron and even smaller for the oxides adjacent. To overcome the problem, so called 'site specific specimen preparation' has been developed since the late 90's using focused ion beam (FIB) technique [35, 36]. The use of gallium ions in a focused ion beam accelerated in a FIB apparatus to energies up to 30 keV enables us to mill specimens selectively to reveal structural features and to deposit films at selected locations. An example carried out in the University of Birmingham UK (2009) is shown in Fig. 23. The specimen is an advanced Ni based superalloy after an interrupted crack propagation test at 650°C in air. A plan-view crack tip TEM specimen was prepared [37]. Using the two different modes: the tungsten deposition to protect the region of interest (Fig. 23 a)) and the milling (trenching) (Fig. 23 b)), the crack tip was transported to a TEM copper grid (c) and d)). The size of the focused Ga ion beam can be achieved as small as a few nanometres in radius. It is possible to prepare the site specific TEM specimen foil as thin as 50 nm in thickness.

Fig. 24 shows TEM bright field images; they are from the same material: polycrystalline advanced superalloy, but they are after different testing conditions. Fig. 24 (a) is from a specimen after interrupting a crack propagation test at 650°C; the intergranular crack propagation was identified. Fig. 24 (b) is from a specimen after interrupting the same crack propagation test mentioned above except for the crack growth rate: 0 μm/s (~0.9 K_{th}: just below the crack growth threshold) held for 5 hours. The oxides ahead of the crack tip are also along grain boundaries. The dashed line boxes indicate the area analysed by EDX shown later in this section. Comparing the two bright field images in Fig. 24, it is apparent that the oxides penetration in the metal ahead of the crack tip is approximately 5 times longer for the static crack specimen (b).

The EDX mapping and the EDX line scan of the oxides close to the crack tip for the moving crack specimen from the region I in Fig. 24 are shown in Fig. 25. It is apparent that the grain boundary is completely filled with oxides (oxygen map). There is a Co and Ni rich oxide in the middle. There are Cr rich oxide areas on both sides of the oxide. Cr thus forms a thin layer between the Ni (Co) oxides and the alloy. EDX line scans across the oxide revealed that Ti, Al are also segregating in this region. Crystallographic analysis using selected aperture diffraction confirmed that the middle oxide is cubic $(Ni_x Co_{1-x})O$ and the rim oxide is hcp $(Cr\ Al\ Ti)_2O_3$. The higher oxygen partial pressure region in the middle of the oxide is consisted by the Ni and Co oxide. The rim of the oxide between the Ni (Co) oxide and the matrix are consisted by the passive Cr, Al and Ti layer. This is correlating with the freshly exposed Ni alloy oxidation process described above. Fig. 26 shows the oxide chemistry at the tip of the oxide corresponding to the region II in Fig. 24. According to the EDX mapping, the oxides formation manner looks similar to that of the region I; Ni(Co) oxide in the middle and the Cr, Al and Ti oxide in the rim. It is questionable to argue the stoichiometry of the oxide from the results of the EDX cross section line scan due to the x-ray emission from the matrix, however it revealed from the line scan in Fig. 26 that the oxide is Cr based; Ni, Co, Al and Ti deplete in the very tip of the oxide. Fig. 27 shows the chemistry of the oxides in the middle of the oxide ahead of the crack tip for the static crack corresponding to the region **x** in Fig. 24. EDX mapping revealed that the Ni and Co are depleted even in the middle of the oxide and the

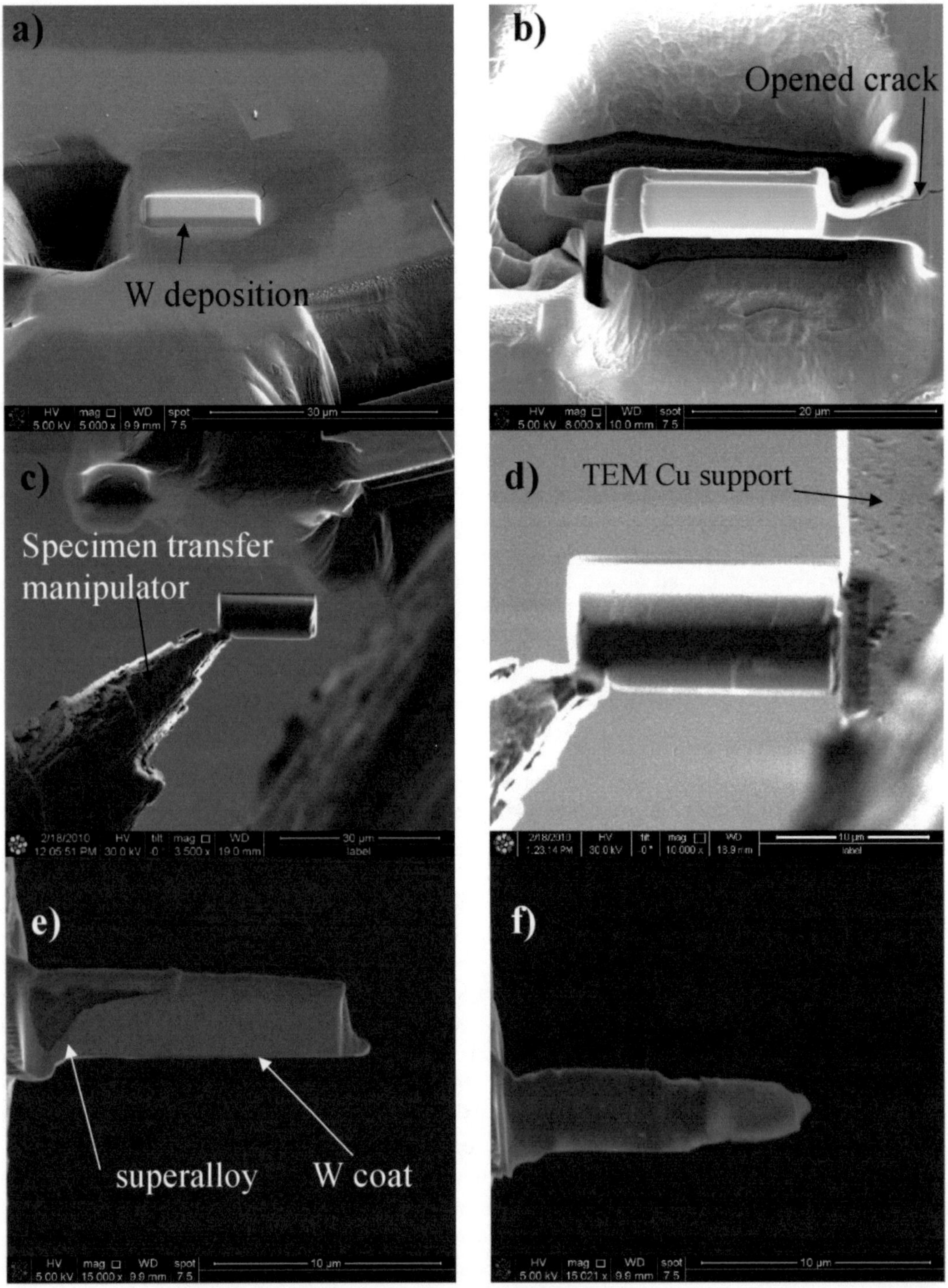

Figure 23. SEM secondary electron images of the process of the site specific TEM sample preparation

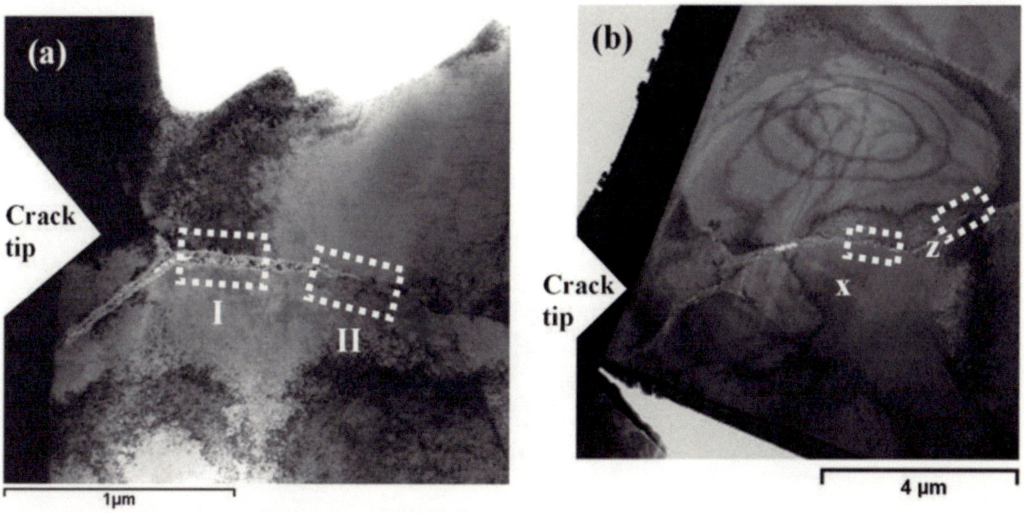

Figure 24. TEM bright field images of the two specimens. The crack tips locate just next to the specimen and the crack propagate from the left to right.

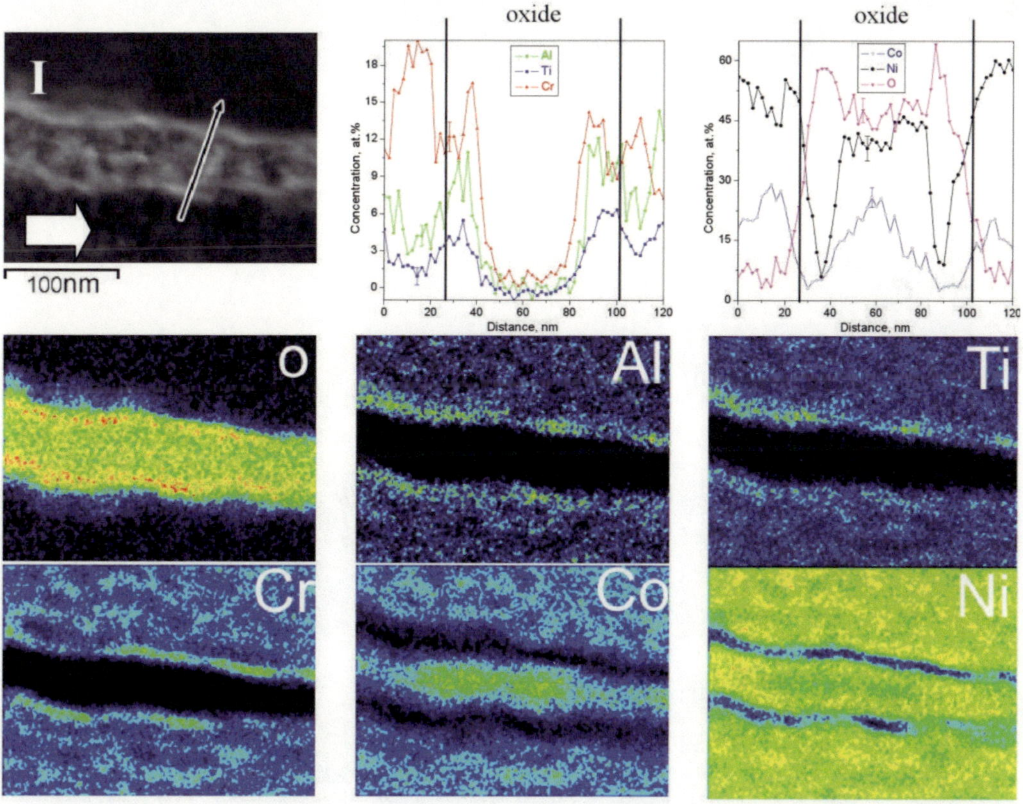

Figure 25. STEM dark field image from the region I in Fig. 24 (top left); the thick white arrow indicates the crack growth direction and the thin black arrow indicates the area and the direction of the EDX line scan. The EDX line scan across the oxide (top right) and the EDX mapping results of the oxide (bottom).

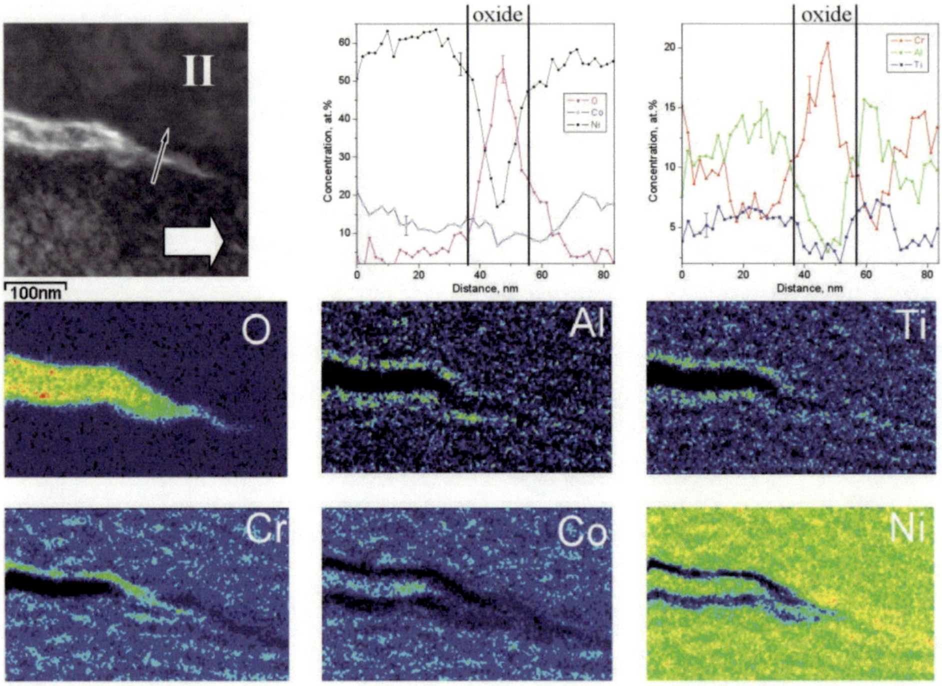

Figure 26. STEM dark field image from the region II in Fig. 24 (top left). The EDX line scan across the oxide (top right) and the EDX mapping results of the oxide (bottom).

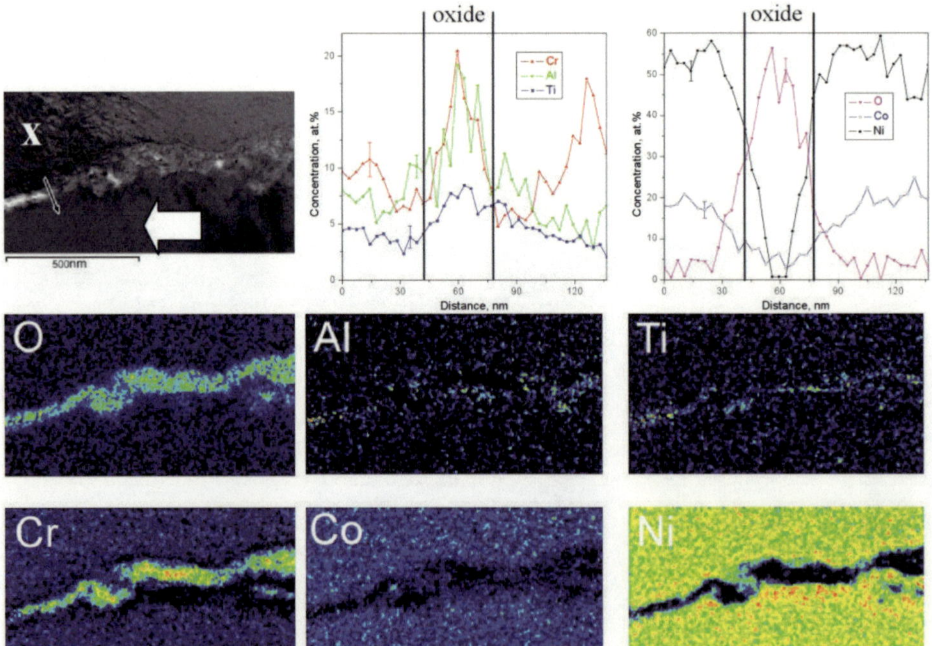

Figure 27. STEM dark field image from the region x in Fig. 24 (top left). The EDX line scan across the oxide (top right) and the EDX mapping results of the oxide (bottom).

line scan revealed that the oxide is consisted mainly by Cr, Al. Fig. 28 shows the tip of the oxide for the static crack corresponding to the region z in Fig. 24. Cr still exists in the oxide in the middle, but particularly at the very tip of the oxide approximately 100 nm or so, Cr is depleted and only Al and Ti enriched at the tip of the oxide (see also the cross section line scan). It is also difficult in this case due to the thickness effect to discuss the stoichiometry of the oxide, however, in this region, Al and Ti based oxide formation takes place.

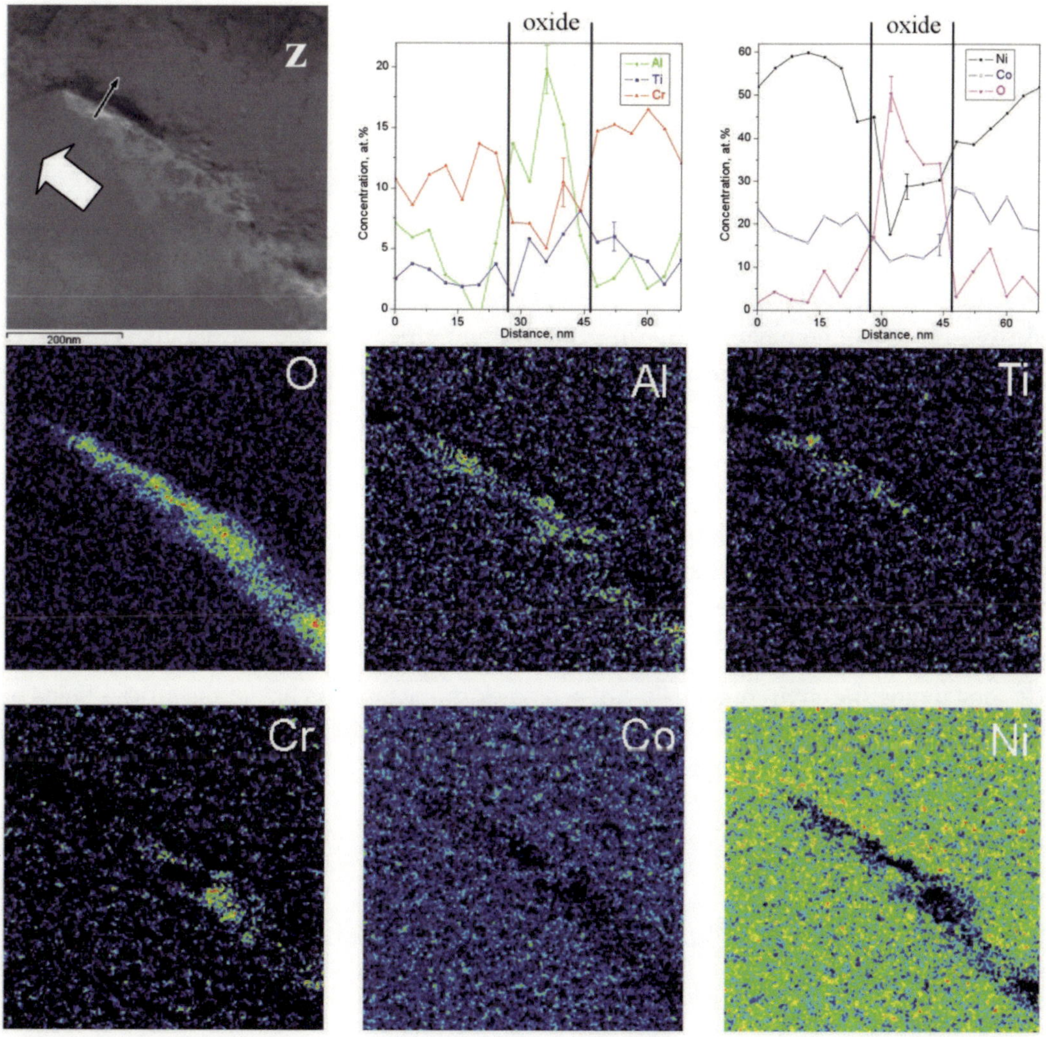

Figure 28. STEM dark field image from the region z in Fig. 24 (top left). The EDX line scan across the oxide(top right) and the EDX mapping results of the oxide (bottom).

Table 2 shows the oxygen dissociation pressure to form the oxide and the free energy for the elements forming oxides introduced above. It is clear that the formation of the oxide at the crack tip follows the thermodynamics suggesting the oxygen partial pressure gradient along the oxide tip ahead of the crack tip.

	Oxide formation	Free energy [kJ/mole]	Oxygen dissociation pressure in log_{10} [bar]
Al	Al2O3	-1425	-48
Ti	Ti2O3	-1295	-44
	TiO2	-798	-40
Cr	Cr2O3	-877	-31
Co	CoO	-166	-18
Ni	NiO	-149	-16

Table 2. Oxide formation free energy and the dissociation pressure at 727°C for the elements introduced in this section [38]

6. Summary

One of the challenges of the advanced Ni based superalloys is in the damage tolerance properties without reducing their superior strength at high temperature. The microstructure, particularly the grain boundaries, was found to be controlled by the two factors in this study. Firstly, it is the nominal chemical composition, especially the hardening precipitate participants. Secondly, heat treatment has a profound influence of microstructure.

The damage tolerance properties are also concerned with the environmentally assisted crack propagation along grain boundaries, which is essentially the oxidation assisted crack propagation in this study. In general, chromium oxide (Cr_2O_3) has been regarded as a healing agent of the oxidation process in Ni alloys. The state-of-the-art technique enabled us to observe the crack tip oxidation. In this study, it was successfully presented that the oxidation sequence is following the free energies for the oxides to form. Thus, for example, Cr_2O_3 is one of the earliest oxides to form at the crack tip.

Understanding the environmentally assisted crack propagation is one of the crucial fields of research to increase the turbine entry temperature (TET), which is indeed one of the most significant attainments in the propulsion industries.

Author details

Hiroto Kitaguchi
Department of Materials, University of Oxford, OX1 3PH, Oxford, UK

Acknowledgement

The author would thank to Profs. I.P. Jones and P. Bowen at The University of Birmingham and Dr. M.C. Hardy at Rolls-Royce plc for their advice and the guidance. The joint support of the UK Engineering and Physical Sciences Research Council (EPSRC) and Rolls-Royce plc is also gratefully acknowledged. Many thanks are also due to Dr H.Y. Li, Research Fellow at the School of Metallurgy and Materials, The University of Birmingham and Dr Y.Y. Tse, former Research Fellow at the School of Metallurgy and Materials, The University of

Birmingham, currently Lecturer at Loughborough University, for their assistance with the mechanical tests and the FIB TEM sample preparation.

7. References

[1] G.W. Meetham, The Development of Gas Turbine Materials 1st ed., Applied Science, London, 1981.
[2] R.C. Reed, The Superalloys, Cambridge University Press, Cambridge, 2006.
[3] H. Cohen, G.F.C. Rodgers, H.I.H. Saravanamuttoo, Gas turbine theory. Third edition, 1987.
[4] C.T. Sims and W.C. Hagel, The Superalloys, John Wiley and Sons Inc., Canada, 1972.
[5] L.R. Liu, T. Jin, N.R. Zhao, X.F. Sun, H.R. Guan, Z.Q. Hu, Materials Science and Engineering A (Structural Materials: Properties, Microstructure and Processing), A361 (2003) 191-197.
[6] X.Z. Qin, J.T. Guo, C. Yuan, J.S. Hou, H.Q. Ye, Materials Letters, 62 (2008) 2275-2278.
[7] S.T. Wlodek, M. Kelly, D.A. Alden, in, TMS, Warrendale, PA, USA, 1996, pp. 129-136.
[8] T. Sugui, W. Minggang, L. Tang, Q. Benjiang, X. Jun, Materials Science and Engineering A, 527 (2010) 5444-5451.
[9] R.J. Mitchell, C.M.F. Rae, S. Tin, Materials Science and Technology, 21 (2005) 125-132.
[10] P.Villars, ASM International, 1997.
[11] K.R. Bain, M.L. Gambone, J.M. Hyzak, M.C. Thomas, Proceedings of the International Symposium on Superalloys, (1988) 13-22.
[12] D. Locq, P. Caron, S. Raujol, F. Pettinari-Sturmel, A. Coujou, N. Clement, Superalloys 2004. Proceedings of the Tenth International Symposium on Superalloys (2004) 179-187.
[13] P.R. Bhowal, E.R. Wright, E.L. Raymond, Metallurgical Transactions A (Physical Metallurgy and Materials Science), 21A (1990) 1709-1717.
[14] N.J. Hide, M.B. Henderson, P.A.S. Reed, SUPERALLOYS 2000. Proceedings of the Ninth International Symposium on Superalloys, (2000) 495-503.
[15] J. Telesman, P. Kantzos, J. Gayda, P.J. Bonacuse, A. Prescenzi, Superalloys 2004. Proceedings of the Tenth International Symposium on Superalloys (2004) 215-224.
[16] J. Telesman, T.P. Gabb, A. Garg, P. Bonacuse, J. Gayda, Superalloys 2008, (2008) 807-816.
[17] E. Andrieu, R. Molins, H. Ghonem, A. Pineau, Materials Science and Engineering A, A154 (1992) 21-28.
[18] H. Ghonem, D. Zheng, Materials Science and Engineering A (Structural Materials: Properties, Microstructure and Processing), A150 (1992) 151-160.
[19] M. Gao, D.J. Dwyer, R.P. Wei, Scripta Metallurgica et Materialia, 32 (1995) 1169-1174.
[20] A.K. Koul, R. Thamburaj, Metallurgical Transactions A (Physical Metallurgy and Materials Science), 16A (1985) 17-26.
[21] D. Turan, D. Hunt, D.M. Knowles, Materials Science and Technology, 23 (2007) 183-188.
[22] E. Andrieu, A. Pineau, Journal de Physique IV (Proceedings), 9 (1999) 3-11.
[23] R. Molins, J.C. Chassaigne, E. Andrieu, Materials Science Forum, 251-254 (1997) 445-452.
[24] R. Molins, G. Hochstetter, J.C. Chassaigne, E. Andrieu, Acta Materialia, 45 (1997) 663-674.

[25] R.G. Baker, J. Nutting, Iron and Steel, 32 (1959) 606-612.
[26] G. Thomas, J. Nutting, Institute of Metals -- Journal, 86 (1957) 7-14.
[27] The Journal of the institute of metals; Discussion, 91 (1963-64) 153.
[28] H. Gleiter, E. Hornbogen, Materials science and engineering, 2 (1968) 285-302.
[29] E. Grant, A. Porter, B. Ralph, Journal of Materials Science, 19 (1984) 3554-3573.
[30] A. Porter, B. Ralph, Journal of Materials Science, 16 (1981) 707-713.
[31] R.D. Doherty, Metal Science, 16 (1982) 1-13.
[32] G.W. Lorimer, R.B. Hicholson, Acta Metallurgica, 14 (1966) 1636.
[33] G.C. Wood, F.H. Stott, Materials Science and Technology, 3 (1986) 519-530.
[34] L.L. Shreir and R.A. Jarman and G.T. Burstein, Corrosion, Butterworth-Heinemann Ltd, Oxford, 1963.
[35] S. Lozano-Perez, Y. Huang, R. Langford, J.M. Titchmarsh, Electron Microscopy and Analysis 2001. Proceedings, 5-7 Sept. 2001, IOP Publishing, Bristol, UK, 2001, pp. 191-194.
[36] D.M. Longo, J.M. Howe, W.C. Johnson, Ultramicroscopy, 80 (1999) 69-84.
[37] F.A. Stevie, R.B. Irwin, T.L. Shofner, S.R. Brown, J.L. Drown, L.A. Giannuzzi, Characterization and Metrology for ULSI Technology. 1998 International Conference, 23-27 March 1998, AIP, USA, 1998, pp. 868-872.
[38] G.V. Samsonov, The oxide handbook, Plenim Publishing Company, Ltd, London, 1973, pp. 23.

Chapter 3

Gold Nanostructures Prepared on Solid Surface

Jakub Siegel, Ondřej Kvítek, Zdeňka Kolská, Petr Slepička and Václav Švorčík

Additional information is available at the end of the chapter

http://dx.doi.org/10.5772/51617

1. Introduction

Up to now, many efforts have been made to produce smart materials with extraordinary properties usable in broad range of technological applications. In particular, within the last two decades, it has been demonstrated that properties of new prospective materials depend not only on their chemical composition but also on the dimensions of their building blocks which may consist of common materials [1,2].

A nanoparticle consists of a few atoms forming a cluster with size in the nanometer range. A nanometer represents a magical size of matter around which the vast majority of materials possess extraordinary, novel physico-chemical properties compared to its bulk form. Considerable attention has been focused during the last few decades on developing and optimizing methods for the preparation of gold nanoparticles to size and shape. Especially properties of 0D spherical and non-spherical particles, as applications of nanostructured materials, may differ considerably depending on the particle shape itself. Simple and straightforward example of the shape dependent behaviour of nanometer-sized particles is its colour. Ultrasmall gold spheres or clusters has been known for centuries as the deep red ruby colour of stained glass windows in cathedrals and domestic glassware. The colour results from the plasmon resonances in the metal cluster. Nowadays, most gold nanoparticles are produced via wet, chemical routes. Nevertheless, synthesis of metal nanoparticles (NPs) has been extensively studied since early 80's [3-9]. Some pioneering works on synthesis of gold nanoparticles were even published as far back as in early 50's by Turkevich [6]. Since that, many techniques have been developed, however, predominately based on wet, chemical processes [4-9]. Currently the most common noble-metal nanoparticle synthesis techniques are those developed by Brust-Shiffrin in 1994 [5]. The method based on reduction of Au^{III+} complex compound with $NaBH_4$ stabilized by thiols enables preparation of high stable particles with pretty narrow distribution and well-controlled size around 1 nm.

Besides interesting properties of nanostructured gold systems such as catalytic effects or magnetism [2,10], which both originate from surface and quantum size effects, they are also extremely usable those, which are closely connected with the average number of atoms in the nanoparticles. The properties and behavior of extremely small gold particles completely differ from those of bulk materials, e.g., their melting point [2,11,12], density [13], lattice parameter [13-15], and electrical or optical properties [13,14,16] are dramatically changed. Gold is also critical component in certain therapies, more specifically, in the treatment of cancer by hyperthermia and thermoablation. These two therapies use heat to kill cancer cells. In the case of hyperthermia, the cancerous tissue is heated to enhance conventional radiation and chemotherapy treatments, while in thermoablation the tissue is heated so that the cancer tissue is destroyed by the localised heat. In principle, there are two methods which can be used to provide heating, i.e. infra-red absorption and the application of an oscillating magnetic field to magnetic nanoparticles. Owing to this, nanosized gold is nowadays used in a vast range of cancer therapy applications such as cancer therapy agents [17] or cancer cell imaging [18,19]. Moreover, gold nanoparticles have often been conjugated with antibodies [20], or grafted to other carriers for surface property enhancement [21,22].

Besides above mentioned 0D nanostructures (nanodiscs, nanoparticles, nanoclusters) increasing efforts have been recently devoted also to one-dimensional (1D) nanostructures. 1D nanostructures in the form of wires, rods, belts and tubes have long been the focus of intensive research owing to their unique applications in mesoscopic physics and fabrication of nanoscale devices [23-25]. It is generally accepted that 1D nanostructures provide a good system for the investigation of the dependence of electrical and thermal transport or mechanical properties on dimensionality and size reduction (quantum confinement) [26]. Of the many elements and compounds from which nanowires may be made, gold is technologically important for its low electrical resistivity (2.21 $\mu\Omega$ cm) [27], its inertness to attack by air and its resistance to sulfur-based tarnishing [28]. Additionally, gold is more biocompatible than most metals, rendering it suitable for implantation [29,30] or electrical interfacing with cells [31,32] and tissues in nanobiological applications [33-35].

Nanostructured materials with high aspect ratios such as nanorods, nanowires, and nanoline patterns often exhibit anisotropic electronic and optical properties that differ from those observed in the bulk materials. These unique materials can be used to create many interesting devices, including fast responding chemical and biochemical sensors [36-40]. The high aspect ratio of nanowires should also make them interesting for the use in two dimensional photonic crystals, where vertical nanowires would constitute an array of high refractive index pillars in air [41]. Field emission from nanowires has also been reported [42], suggesting the possibility of devices such as field emission displays (FEDs) with nanowires acting as cathodes.

A variety of fabrication techniques have been developed in the past decade that yield high quality nanowires. Fabrication of ordered arrays of metallic nanoparticles supported on transparent substrate by sequential techniques like electron beam lithography has been demonstrated [43]. Such top-down approaches, however, are cumbersome and have a low yield, which hinders practical applications. High throughput approaches for the synthesis of metallic nanowires are thus intensely searched [44-48]. In general, the production of arrays of nanostructures on substrates by lithographic techniques presents the disadvantage of high cost and a restriction in the number of materials to which it can be applied. The method can also prove to be complex and inefficient. Template based methods overcome those disadvantages, but the obtained structures often present a high number of imperfections due to packing defects in the original templates [49].

Above mentioned applications, however, usually require gold nanostructures (0D or 1D) to be either suspended in colloid solution or attached to another support medium. Concerning this, creation of nanostructured gold directly on appropriate support may be technologically valuable since one can avoid additional preparation step oriented on metal-substrate mutual attachment. Therefore, this chapter focuses on new possible approaches for nanostructuring of gold layers either formerly deposited on solid substrates (polymer or glass) or during deposition itself (polymer). The formerly mentioned technique is based on the intensive post-deposition thermal annealing of sputtered layers on polytertaflouroethylene (PTFE) or glass, whereas the latter technique is based on forced (preferential) growth of gold on nanostructured polymer template. The method, combining nanoscale patterning of the polyethyleneterephtalate (PET) substrate by polarized light of excimer laser with glancing angle deposition of the gold, provides an interesting alternative to time consuming sequential lithography-based nanopatterning approaches.

First part of the text is focused on the study of selected physico-chemical properties of deposited gold layers and its changes induced by post-deposition annealing. The gold nanostructures of different thicknesses were sputtered onto glass or polymer (PTFE) substrate and then the samples were annealed from room temperature to 300°C. The effects of annealing on gold structures sputtered onto substrate, their surface morphology and roughness were studied using Atomic Force Microscopy (AFM), lattice parameter and crystallites size and their distribution by X-ray diffraction (XRD) and by SAXSess. Hall mobility, volume resistance and free carrier concentration were measured by Van der Pauw method, an electric permitivity by ellipsometry, an optical band gap by UV-Vis spectroscopy and a sheet resistance of gold nanostructures by 2-point method.

In the next part special attention will be given to the irradiation of PET surface with linearly polarized light of a pulsed KrF excimer laser to produce templates for preparation of laterally ordered self-organized arrays of metallic nanowires. Different fluences and angles of incidence of the laser beam were applied. The periodicity of the ripples created on the polymer surface was controlled by changing the incidence angle of laser light during irradiation. Subsequently the modified polymer surface was coated with gold using two deposition techniques (sputtering and evaporation). The surface of nano-patterned coated/uncoated PET was analyzed by AFM and a scanning electron microscopy equipped with a focused ion beam (FIB-SEM), allowing to cut cross-sections of the laser patterned substrate surface and the deposited gold layers

2. Gold nanostructures on glass substrate

An overview of growth process, morphology, electrical and optical properties of ultra-thin gold layers sputtered on glass is provided in following sections. Insight into the phenomena taking place during post-deposition thermal treatment is also given.

2.1. Thickness, morphology and inner structure

Thickness of sputtered layers was measured by AFM. Thickness in the initial stage of deposition (sputtering time less than 50 s) was determined from the SEM image of the sample cross-section (FIB-SEM). Dependence of the layer thickness on sputtering time is shown in Fig. 1. Linear dependence between sputtering time and structure thickness is evident even in the initial stage of the layer growth. This finding is in contradiction with results obtained earlier for Au sputtering on PET [50]. In that case, the initial stage of the layer growth was related to lower deposition rate which is due to different morphology.

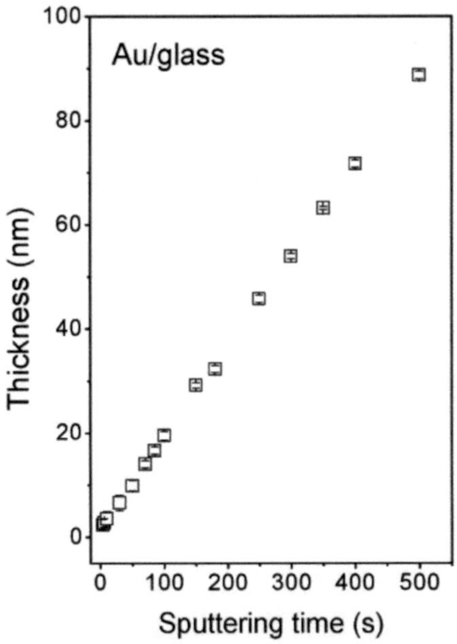

Figure 1. Dependence of the gold structure thickness on sputtering time [13].

In Fig. 2, a SEM picture of the cross-section of the Au layer at its initial stage of growth is shown. It is obvious that after approximately 20 s of Au deposition, flat, discrete Au islands (clusters) appear on the substrate surface. The flatness may indicate preferential growth of gold clusters in a lateral direction. When the surface coverage increases and the clusters get in close contact with each other, a coarsening sets in and becomes the dominant process. After the surface is fully covered, additional adsorption causes only the vertical layer growth, while the lateral growth is dominated by cluster boundary motion [51].

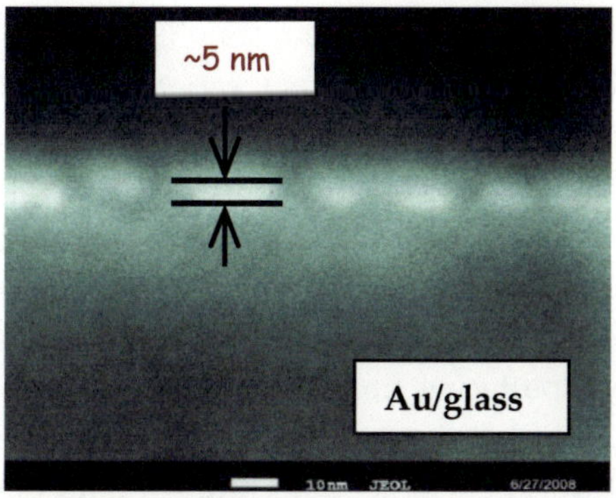

Figure 2. SEM scan of the FIB section of gold structure on glass substrate. Deposition time was 20 s. [13].

The AFM images that illustrate the surface morphology and roughness (R_a) of gold-coated glass before and after annealing are shown in Fig. 3. For the sake of comparison only images of the samples with identical vertical scale were chosen. From Fig. 3 it is clear that the surface morphology of the as-sputtered structures does not depend significantly on the sputtering times. Monotonous decrease of surface roughness with deposition time is related to the stage of the layer growth. During initial stages of metal growth the layer is formed over isolated islands. After that, during ongoing deposition, interconnections between clusters are formed and the deposited layers become homogeneous and uniform. Decrease of surface roughness is direct evidence of the formation of a thicker layer during sputtering process on flat substrate. After annealing, however, the surface morphology changes dramatically. Similar changes in the morphology of the thin gold structures have also been observed on the samples annealed at 200°C for 20 hours [52] and at 450°C for 2 hours [11]. It is seen from Fig. 3 that the annealing leads to the formation of "spherolytic and hummock-like" structures in the gold layers. The formation may be connected with an enhanced diffusion of gold particles at elevated temperature and their aggregation into larger structures. It is well known that the melting point of the gold nanoparticles decreases rapidly with decreasing particle size [2,11,12].

The migration of the gold nanoparticles and formation of larger structures may be connected with lower thermodynamic stability of the gold nanoparticles and lower gold wettability of glass. This idea is supported by some previous XRD experiments in which dominant (111) orientation of gold crystals in the sputtered gold layers was determined [11, 53]. The (111) oriented gold crystals are known to be thermodynamically unstable and their melting and cracking starts from the edge parts that should be bounded to Au (110) surface [11].

Metallic nanoparticles and generally nanostructures composed of metals often exhibit different values of structure parameters compared to their bulk form e.g. contraction of lattice parameter in nanostructures increases material density [13,53]. Lattice parameters *a* of the face-centered cubic gold nanostructures determined before and after annealing are shown in Fig. 4 as a function of the sputtering time (i.e. effective layer thickness). Lattice parameters were calculated using the Rietveld procedure (full pattern fitting). For this purpose the five strongest diffraction lines were taken into account. For very thin films the diffraction lines are weak and the resulting values of the lattice parameters are loaded by a higher error. The error is especially large for the as-sputtered samples. A dramatic difference is found in the dependences of the lattice parameter on the sputtering time between as-sputtered and annealed samples. For as-sputtered samples the lattice parameter varies rapidly with the increasing sputtering time, i.e. with the increasing mean size of the gold crystallites [16,53,54]. It is seen that a maximum lattice parameter is observed after 100 s of sputtering, i.e. for the layer thickness of about 18 nm. For both the thinner and thicker layers the lattice parameter declines significantly. The same trend in the lattice parameter vs. sputtering time dependence was reported also for silver structures, for which the lattice parameter increases slightly up to the structures size of 12 nm and then decreases [55]. In contrast to the as-sputtered gold structures at the annealed ones the lattice parameter is nearly independent on the sputtering time and the size of the structures.

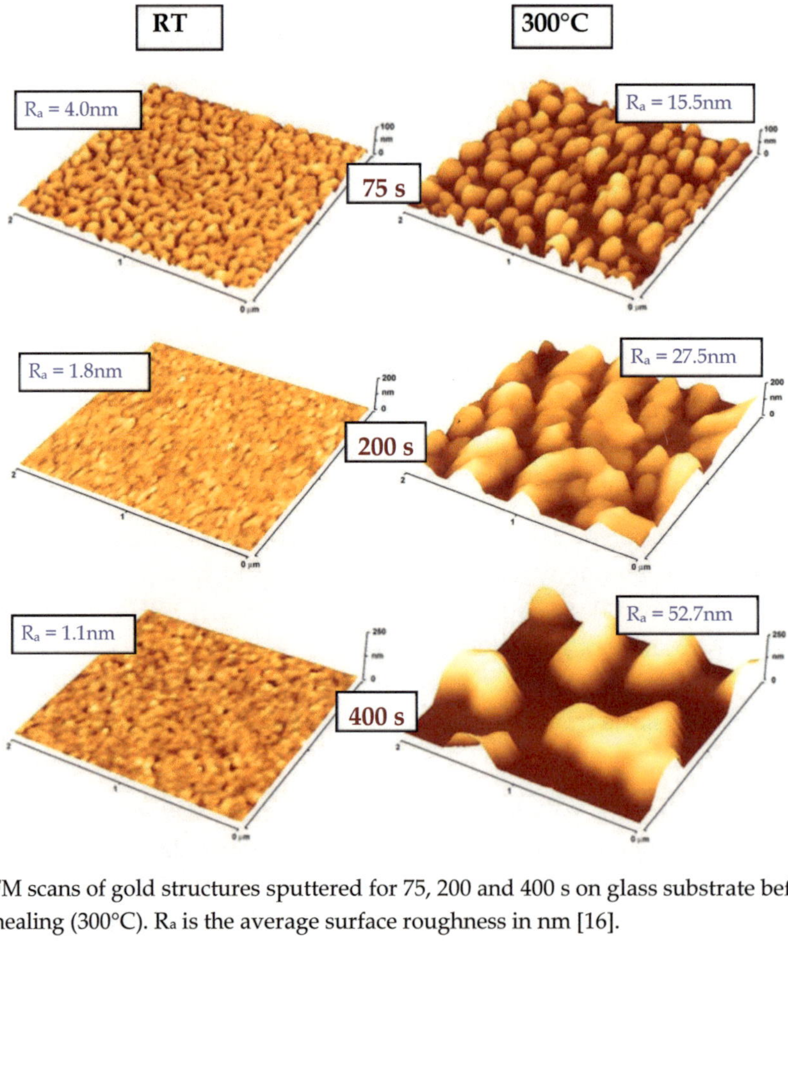

Figure 3. AFM scans of gold structures sputtered for 75, 200 and 400 s on glass substrate before (RT) and after annealing (300°C). R_a is the average surface roughness in nm [16].

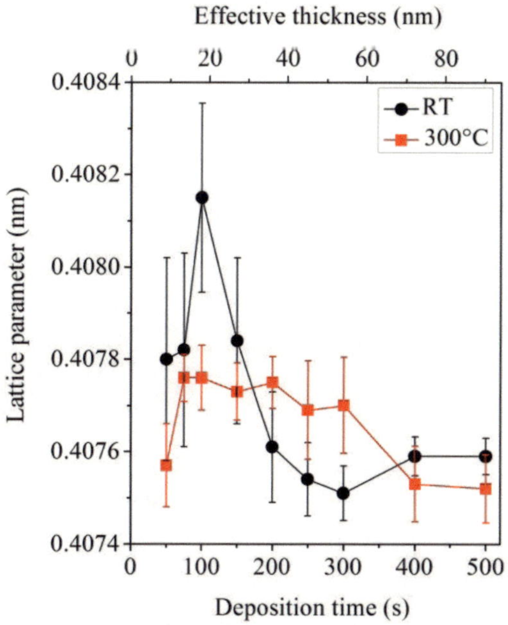

Figure 4. Dependence of the gold lattice parameter on the sputtering time (i.e. effective thickness) measured before (-•-) and after annealing at 300°C (-■)[14].

The dependence of the crystallite size on the sputtering time before and after annealing is shown in Fig. 5. The dependences are quite different for the as-sputtered and annealed samples. While in the as-sputtered samples the crystallite size is amonotonously increasing function of the sputtering time, in the annealed ones the crystallite size first increases rapidly up to the sputtering time of 250 s, achieves a maximum and then it decreases. A dramatic increase of the Au crystallite size with the annealing temperature was published also by Santos et al. [56].

Size distribution of the Au crystallites determined by SAXSess method is presented in Fig. 6. The mean crystallite size values (modus) determined by SAXSess (S) and by XRD (X) are compared in Fig. 5. It is obvious that before annealing both methods (SAXSess, XRD) give the same values of the mean crystallite size, which increase slightly with the deposition time. Both SAXSess and XRD measurements prove dramatic increase of the mean crystallite size after annealing. However, there is an obvious dissimilarity between SAXSess and XRD results regarding longer sputtering time which is caused by inability of the SAXSess method to examine crystallites larger than ca 90–100 nm. The different behavior may be due to a crystallites' re-crystallization in the annealing process. The crystallite size determined by the XRD technique is based on the determination of the so-called coherently diffracting domains with their mean dimensions in direction perpendicular to the film surface. This is the reason why the crystallites determined in this way significantly exceed their size in some cases.

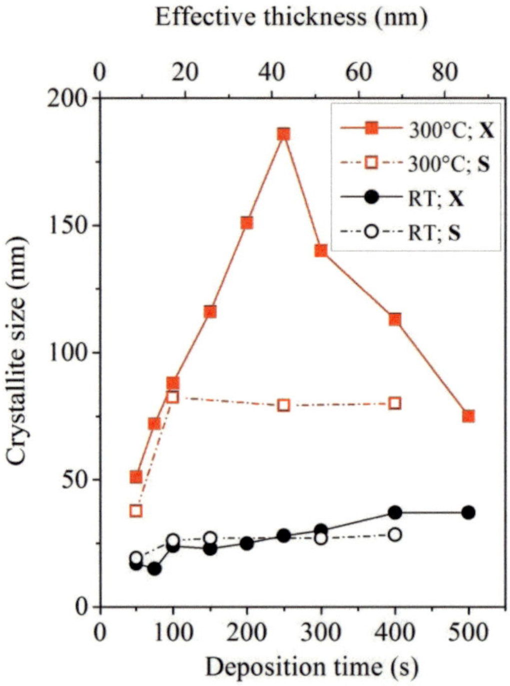

Figure 5. Dependence of the size of the gold crystallites on the sputtering time (i.e. layer effective thickness) measured before (-○S, -•- X) and after annealing at 300°C (-□- S, -■X) using S – SAXSess, X – XRD methods [14].

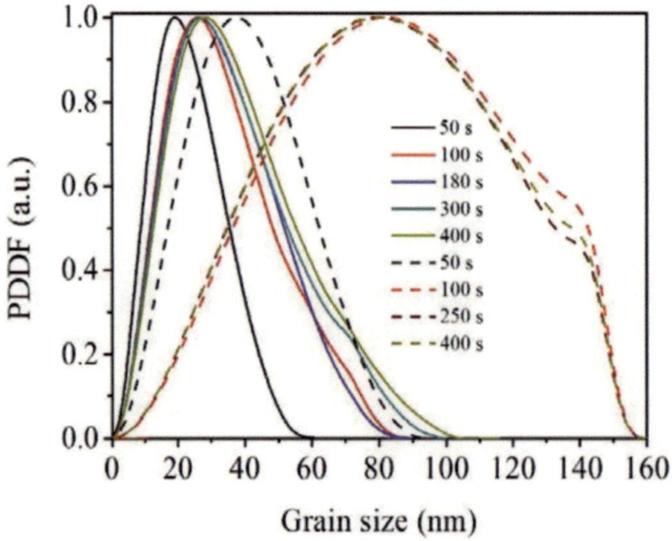

Figure 6. Pair distance distribution functions (PDDF) of gold crystallites by different sputtering time (in seconds) before (solid line) and after annealing (dashed line) measured by SAXSess method [14].

2.2. Optical and electrical properties

Besides interesting catalytic and electronic properties, nanoparticles of noble metals exhibit also distinctive, shape-dependent optical properties that have attracted great technological interest. This is particularly true for gold nanostructures [11]. Images of the sample surface for different sputtering times and for sputtered and annealed samples are shown in Fig. 7.

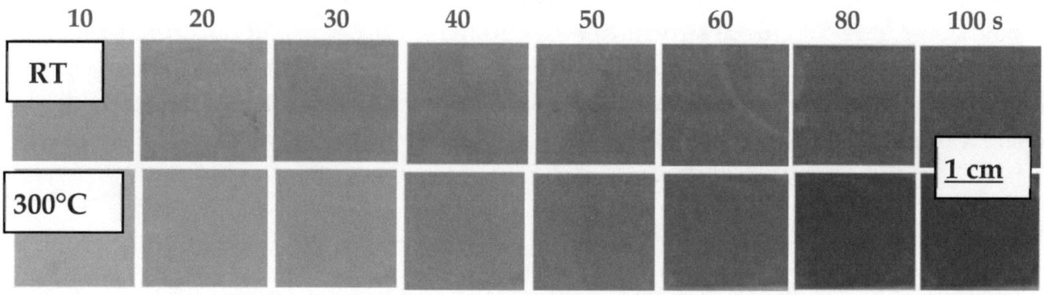

Figure 7. Images of the glass samples with gold structures sputtered for increasing times. The as-sputtered (RT) and annealed samples (300°C) are shown for comparison [16].

The deposited samples become darker with increasing sputtering time, the darkening being related to increasing thickness of the gold structures. Also a gradual change of the structures colour from blue to green is seen. After annealing all structures exhibit reddish colour, regardless of the sputtering time. The changes in the layer colour indicate pronounced alteration in the gold nanostructure caused by the annealing (see Fig. 3). It could be in accordance with previously presented results, the small gold sample about 10 nm absorbs green light and thus appears red [2]. This effect was confirmed also by UV-Vis spectroscopy. For the sake of clarity only some of UV-Vis spectra from as-sputtered and annealed samples are shown in Fig. 8 (RT and 300°C). The absorbance of gold structures increase with increasing sputtering time and structure thickness as could be expected. From comparison of the spectra of the sputtered and annealed samples it is seen that the annealed structures have qualitatively different shapes and lower absorbance. Both phenomena point at structural changes due to annealing. The observed shift of the 530 nm absorption peak (corresponding to surface plasmon resonance) with increasing sputtering time towards longer wavelengths is probably related to interconnection and mutual interaction of gold nanosized islands in the structure. From present UV-Vis spectra it is evident that the as-sputtered samples prepared for deposition time of 30 s and that annealed one (sputtering time 200 s) are the first ones, which do not exhibit the peak of plasmon resonance. Qualitative difference between absorbances of the sputtered structure and that annealed may indicate a transition from the structure comprising discrete gold islands to continuous gold coverage.

The UV-Vis spectra were also interpreted in the frame of the well-known Tauc's model [57] and the optical band gap (E_g^{opt}) was calculated as a function of the sputtering time for sputtered and annealed samples. This dependence is shown in Fig. 9. Also the gold structure thickness vs. sputtering time measured by AFM method on the scratch step is presented. These values, inclusive the result that the AFM-scratch technique is not applicable on annealed structures due to their altered morphology, were taken from our previous work [53]. It is seen from Fig. 9 that the structures sputtered for times below 150 s exhibit non-zero E_g^{opt}. Rather dramatic change in the E_g^{opt} is observed after annealing, where the values of the E_g^{opt} are much higher in comparison with those of the sputtered sample. For samples sputtered for times around 300 s a non-zero E_g^{opt} is observed. For behaviour of ultra-thin metal structures (<10 nm) the surface-size and quantum-size effects must be considered [2,53,58]. This quantum-size effect in small structures leads e.g. to a semi-conducting character, which is accompanied by non-zero E_g (band gap) or E_g^{opt}. This effect was observed in the present case.

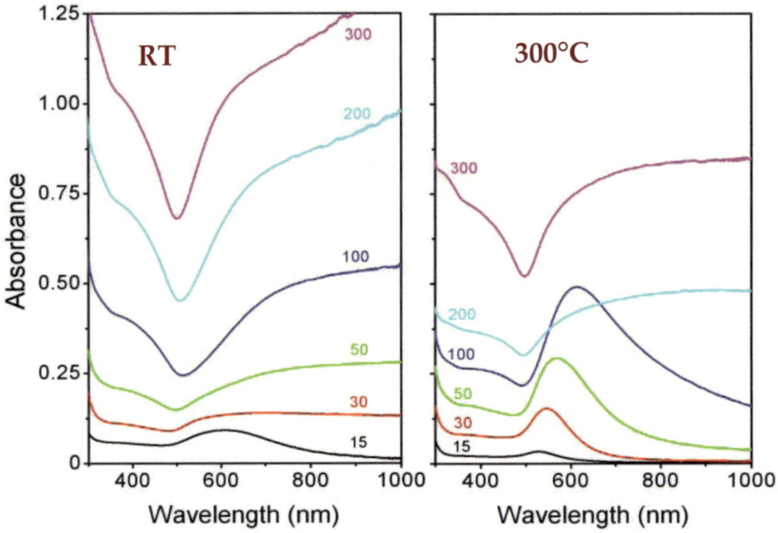

Figure 8. UV-Vis spectra of gold structures sputtered on glass before (RT) and after annealing (300°C). The numbers in Figs. are sputtering times in s [16].

The dependence of the volume resistivity on the sputtering time is seen from Fig. 10a. For as-sputtered samples a rapid drop of the resistivity over a narrow thickness interval is observed. The drop indicates a transition from electrically discontinuous to continuous gold coverage. For annealed samples the resistivity drop is shifted towards thicker layers. The difference is obviously connected with changes in the layer structure taking place during annealing i.e. gold coalescence and formation of isolated islands [16]. The onset of the rapid resistivity drop is observed after 50 and 150 s of sputtering for as-sputtered and annealed samples, respectively. Free carrier volume concentration and their Hall mobility significantly affect the electrical conductance of materials. The dependence of the free carrier concentration and the mobility on the sputtering time is shown in Figs. 10b and 10c, respectively. As can be seen from Fig. 10b, with increasing sputtering time the carrier concentration increases dramatically and the layers become conductive (see Fig. 10a). As in the case of resistivity the onset of the rapid increase of the free carrier concentration on annealed samples is shifted towards longer sputtering time. Thus the constant level of the free carrier concentration is achieved later compared to the as-sputtered samples. A similar dependence of the free carrier concentration on the layer thickness was recently observed on PET and PTFE sputtered with gold [59]. The carrier mobility also changes dramatically with increasing sputtering time for non-annealed and annealed samples (Fig. 10c). The mobility first declines rapidly to a point when an electrically continuous layer is formed. The decline may be due to the fact that in a discontinuous layer the mobility mechanism differs from classical electron conductivity common in metals. For annealed structures the continuous layer is formed after a longer deposition time. For thicker, electrically continuous gold layers the mobility is a slowly increasing function of the sputtering time.

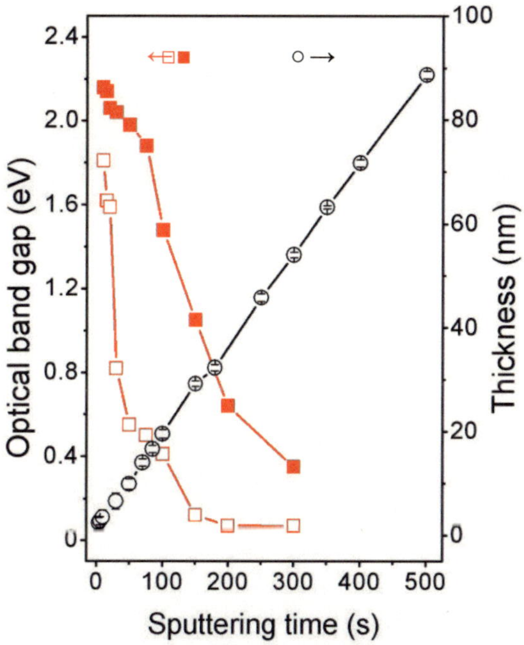

Figure 9. Dependence on the sputtering time of the optical band gap of gold structures before (RT) and after annealing (300°C) (-□- for RT and -■ for 300°C) and thickness (-○) [16].

There is a clear correspondence between mobility (Fig. 10c), free carrier volume concentration (Fig. 10b) and volume resistivity (Fig. 10a). A similar dependence of the free carrier concentration on the thickness of the gold layers deposited by sputtering on PET and PTFE was observed [59]. Simple and straightforward interpretation of the above described observations is that during electrical measurement on discontinuous gold layers an electron injection due to the tunneling effect occurs [13,16]. With ongoing deposition time the discrete structures become interconnected and form an electrically continuous, homogeneous layer in which the concentration of free carriers is saturated.

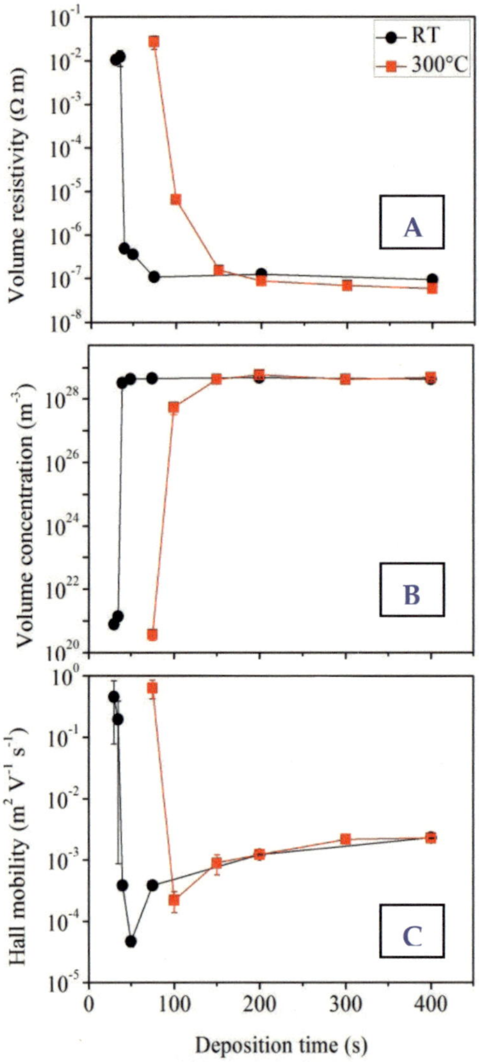

Figure 10. Dependence of the volume resistivity (A), surface free carrier volume concentration (B) and surface free carrier Hall mobility (C) on the sputtering time measured by van der Pauw technique before (-•-) and after annealing at 300°C (-■) [14].

The IR part of optical constants of the as-deposited and annealed Au films determined from ellipsometry also supports the results of electrical transport measurements. Fig. 11 presents the real part of electric permittivity in the studied spectral range.

Spectroscopic features in the Drude (IR) region clearly show the tendency of Au films to lose their metallic behavior with decreasing thickness due to gold coalescence, leading to a layer discontinuity [16]. Film discontinuity of the as-deposited thin layers is a natural consequence of the mechanism of the layer growth. Percolation threshold is reached at the layer thickness of about 7 nm [60] corresponding to the deposition time of about 25 s in the present case. Fig. 11 also shows that the strong change in the surface morphology induced by the annealing shifts the metal-to-insulator transition towards greater layer thicknesses (i.e. deposition times). The thickness variation of IR end of the real part of the electric permittivity spectra of annealed gold layers (positive value reaching the maximum and then passing through zero to negative values with increasing deposition times) is consistent with previous studies of metallic films around the percolation threshold [60]. For the annealed layers with sputtering times equal to or smaller than those corresponding to the metal-to-insulator transition, a strong signature of plasmons is expected in the VIS part of optical constants. This is documented in Fig. 12, where the spectral dependence of the imaginary part of the electric permittivity is shown.

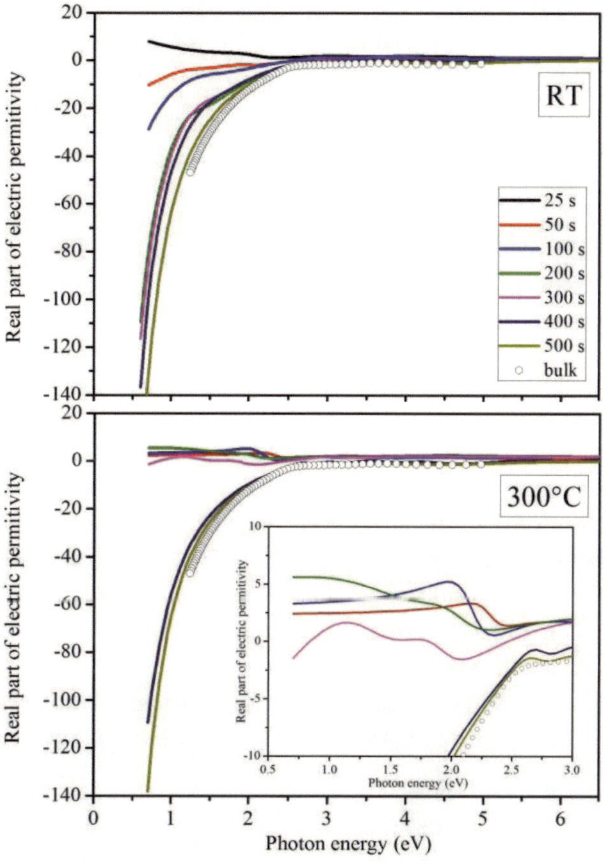

Figure 11. Real part of the electric permittivity spectra of as-sputtered (RT, upper part) and annealed (300°C, lower part) gold structures obtained by spectroscopic ellipsometry for the different sputtering times [14].

Plasmon oscillator band for the layer sputtered for 50 s is centered at around 2.3 eV (540 nm). With increasing deposition time the band becomes broader and shifts to longer wavelengths (lower photon energy). For 200 s sputtering time the plasmon band splits into two and for longer sputtering times it integrates into the Drude term in the IR spectral limit.

This change in the optical constants around the metal-to-insulator transition is the reason for the color variation of the annealed layers.

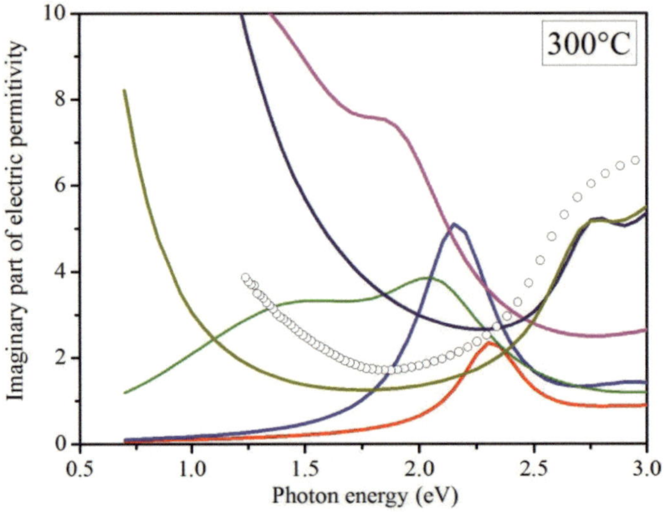

Figure 12. Imaginary part of the electric permitivity spectra of annealed (300°C) gold structures. The presence and evolution of the plasmon bands should be noted (for details see the text). Coloured sign of curves is the same as in Fig. 11 [14].

The temperature dependence of the sheet resistance for two particular structure thicknesses is displayed in Fig. 13. One can see that the temperature dependence of the sheet resistance strongly depends on the structure thickness. For the layer about 89 nm thick, the resistance is an increasing function of the sample temperature, the expected behavior for metals. For the structure about 6 nm thick, the sheet resistance first decreases rapidly with increasing temperature, but above a temperature of about 250 K, a slight increase in resistance is observed. The initial decrease and the final increase of the sheet resistance with increasing temperature are typical of semiconductors and metals, respectively. It has been referred elsewhere [2] that a small metal cluster can exhibit both metal and semiconductor characteristics just by varying the temperature. It is due to temperature-affected evolution of band gap and density of electron states in the systems containing low number of atoms.

From the present experimental data, it may be concluded that for the thicknesses above 10 nm, the sputtered gold layers exhibit metal conductivity. In the thickness range from 5 to 10 nm, the semiconductor-like and metal conductivities are observed at low and high temperatures, respectively. Our further measurements showed that the layers thinner than 5 nm exhibit a semiconductive-like characteristic in the whole investigated temperature scale. Except for band gap evolution theory, typical semiconductor-like behavior may also originate from the tunneling effect of electrons through the discontinuous, separated Au clusters during electrical measurements. Since the probability of electron tunneling depends on the temperature, similarly, typical course of sheet resistance and, as will be shown later, CV characteristic may be affected right by this phenomenon.

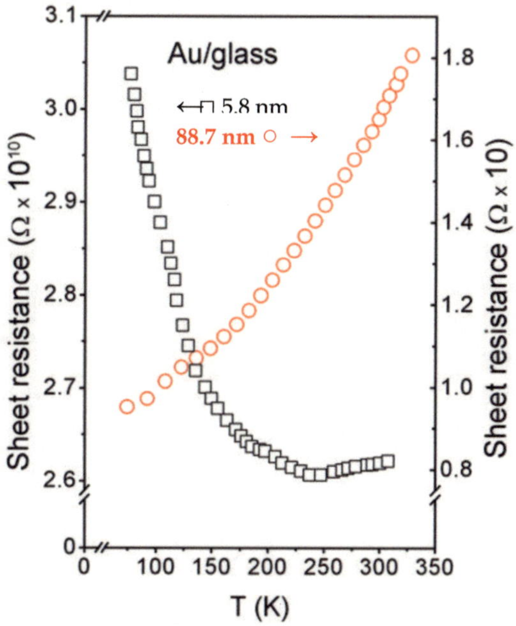

Figure 13. Temperature dependence of the sheet resistance for two different structure thicknesses indicated in the figure [13].

From presented measurements of sheet resistance results the semiconductor-like character of Au at specific structure conditions (thickness, temperature). The observed semiconductor-like character (decreasing resistance with increasing temperature) of ultrathin Au structures may originate from two undistinguishable phenomena. The first one results from a tunneling effect which occurs at discontinuous structures during resistance measurements [59]. The second one originates from the semiconductor characteristic of the intrinsic cluster itself, which occurs in metal nanostructures of sufficiently small proportions [2].

3. Gold nanostructures on polymeric substrate

In this section special attention is given to the changes in surface morphology and other physico-chemical properties of gold nanolayers, sputtered on polytetrafluoroethylene (PTFE) surface induced by post-deposition annealing.

3.1. Electrical properties

The dependence of the electrical sheet resistance (R_s) of the gold layer on its thickness before and after annealing (at 100, 200 and 300°C) is shown in Fig. 14. For the as-sputtered samples the sheet resistance decreases rapidly in the narrow thickness range from 10 to 15 nm when an electrically continuous gold coverage is formed. The resulting sheet resistance is saturated at a level of aproximately 200 Ω. From the measured R_s and effective layer thickness, layer resistivity R (Ohm centimeter) was calculated, which appears to be few orders of magnitude higher than that reported for metallic bulk gold ($R_{Au}^{bulk} = 2.5 \times 10^{-6}$ Ω cm [61], e.g., for 100 nm thick Au layer $R_{Au}^{100\ nm} = 1 \times 10^{-3}$ Ω cm). As in the case of Au coated glass substrate (see section 2), the higher resistivity of thin gold structures is due to the size effect in accord with the Matthiessen rule [62]. Annealing at temperatures below 200°C causes only mild shift in the resistance curve towards thicker layers. Transition from electrically discontinuous to electrically continuous layer in case of low temperature annealed samples is more gradual and occurs between the effective layer thicknesses from 10 to 20 nm regarding the annealing temperature. After annealing at 300°C a dramatic change in the resistance curve is observed. The annealed layers are electrically discontinuous up to the Au effective thickness of 70 nm above which the continuous coverage is created and a percolation limit is overcome. However, for longer sputtering times up to 550 s, the sheet resistance changes slowly and it achieves a saturation which is observed on the as-sputtered layers and layers annealed at low temperatures.

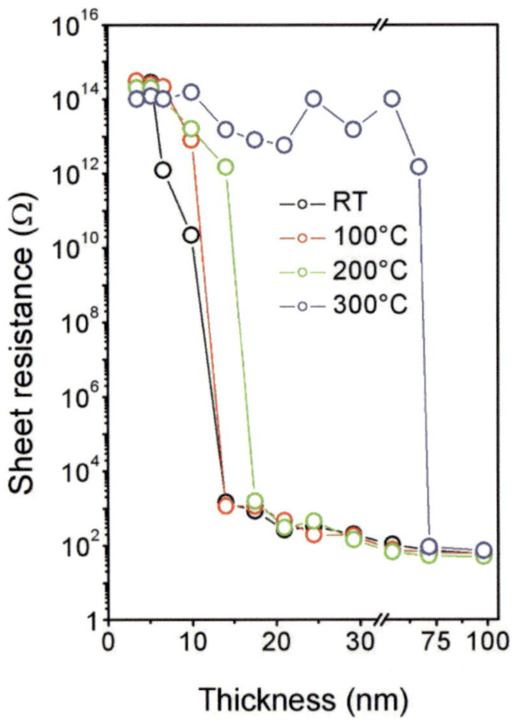

Figure 14. Dependence of the sheet resistance (R_s) on Au layer thickness for as-sputtered samples (RT) and the samples annealed at 100, 200 and 300°C [63].

Compared to electrical properties discussed in chapter 2 (Au layers on glass substrate), one can see that in case of PTFE substrate the transition from electrically discontinuous to continuous layer is shifted towards thicker layers. This fact is due to incomparable value of surface roughness of substrate used which is in the case of PTFE one order of magnitude higher (see section 3.3).

3.2. Chemical composition

Besides the sheet resistance measurements, information on the layer structure and homogeneity can be obtained in another way too. Here, complementary information on the layer homogeneity is obtained from XPS spectra. Fig. 15 A,B shows intensity normalized XPS spectra (line Au 4f) of 20 and 80 nm thick sputtered gold layers, respectively. Black line refers to as-sputtered layer and blue line to the one annealed at 300°C. Annealing of the 80 nm thick gold layer does not change the XPS spectrum. In contrast, the annealing of the 20 nm thick layer results in strong broadening of both lines which is due to the sample charging in the course of the XPS analysis. The charging is closely related to the change in the layer morphology: from electrically continuous one for as-sputtered sample to discontinuous one after the annealing procedure [16]. This observation is in agreement with above described results of the sheet resistance measurements (see Fig. 1, section 3.1).

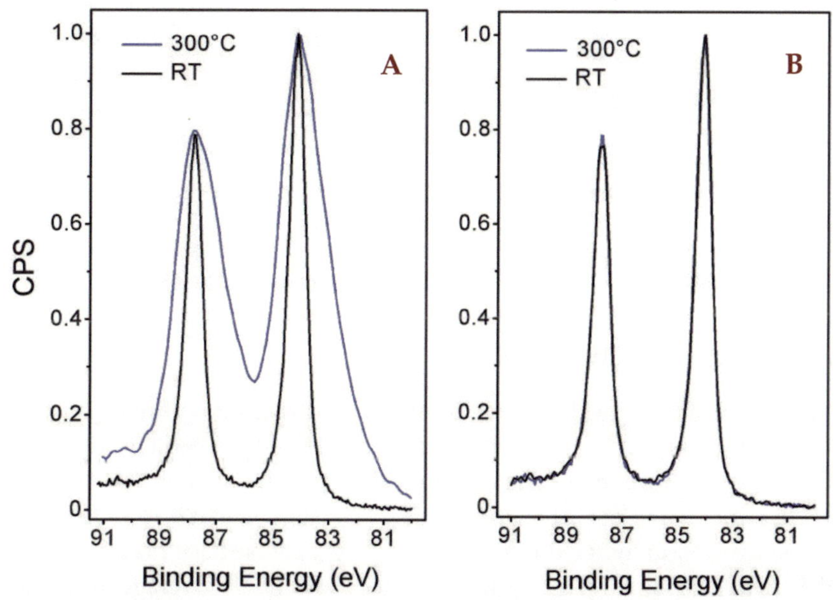

Figure 15. Intensity normalized XPS spectra (line Au (4f)) of 20 (A) and 80 nm (B) thick sputtered Au layers on PTFE before (black line) and after (blue line) annealing at 300°C [63].

Concentrations of chemical elements on the very sample surface (accessible depth of 6 to 8 atomic layers) determined from XPS spectra are summarized in Table 1. The XPS data were obtained for the samples with 20 and 80 nm thick gold layers, both as-sputtered and annealed at 300°C. Total carbon concentration and the carbon concentration coming from PTFE (calculated from XPS data) are shown in columns 1 and 2 of the table, respectively. Major part of the carbon is due to sample contamination. Fluorine to PTFE carbon ratio F/C^{PTFE} is close to that expected for PTFE (about 2). By the annealing at 300°C, the ratio decreases to 1.7 for both layer thicknesses. The decrease may be due to reorientation of polar

C-F groups induced by thermal treatment. Oxygen detected in the samples may result from oxygen incorporation during gold sputtering which may be accompanied by partial degradation and oxidation of PTFE macromolecular chain or degradation products. Subsequent annealing leads to reorientation of the oxidized groups toward the sample bulk and corresponding decrease of the surface concentration of oxygen. The same effects have been observed earlier on plasma-modified polyolefines [64]. It is also evident from Table 1 that annealing causes resorption of contamination carbon both hydrogenated and oxidized one [65]. Changes in the morphology of the gold layer after the annealing are manifested in changes of the gold and fluorine concentrations as observed in XPS spectra. After the annealing, the observed gold concentration decreases and fluorine concentration increases dramatically, these changes clearly indicate formation of isolated Au islands similar to those in case of Au-coated glass substrate [16].

Au layer Thickness	Temperature	Atomic concentrations of elements in at. %					
		C	C^{PTFE}	O	Au	F	F/C^{PTFE}
20 nm	RT	43.5	4.4	6.5	41.6	8.5	1.93
	300°C	37.8	34.8	0.4	3.4	58.4	1.68
80 nm	RT	41.0	3.1	4.4	48.6	6.0	1.94
	300°C	36.8	27.2	1.2	14.8	47.2	1.74

Table 1. Atomic concentrations (in at. %) of C (1s), O (1s), Au (4f) and F(1s) in Au sputtered PTFE samples with Au effective thickness 20 and 80 nm after deposition (RT) a after annealing (300°C) measured by XPS. C^{PTFE} represents calculated concentration from XPS data of carbon (in at. %) originating from PTFE only, F/C^{PTFE} stands for fluorine to PTFE carbon ratio [63].

3.3. Surface properties and morphology

Another quantity characterizing the structure of the sputtered gold layers is zeta potential determined from electrokinetic analysis. Dependence of zeta potential on the gold layer thickness for as-sputtered samples (RT) and annealed samples at 300°C is shown in Fig. 16. For as-sputtered samples and very thin gold layers, the zeta potential is close to that of pristine PTFE due to the discontinuous gold coverage since the PTFE surface plays dominant role in zeta potential value. Then, for thicker layers, where the gold coverage prevails over the original substrate surface, the zeta potential decreases rapidly and for the thicknesses above 20 nm remains nearly unchanged, indicating total coverage of original substrate by gold. For annealed samples, the dependence on the layer thickness is quite different. It is seen that the annealing leads to a significant increase of the zeta potential for very thin layers. This increase may be due to thermal degradation of the PTFE accompanied by production of excessive polar groups on the polymer surface, which plays the important role when the gold coverage is discontinuous. Moreover, the surface roughness increases at this moment too (see Table 1 and Fig. 17 below) [66]. Then, for medium thicknesses, ranging from 20 to 70 nm, the zeta potential remains unchanged and finally it decreases again for higher thicknesses due to the formation of continuous gold coverage. It appears that the results of electrokinetic analysis (Fig. 16) and measurement of the sheet resistance (Fig. 13) are highly correlated.

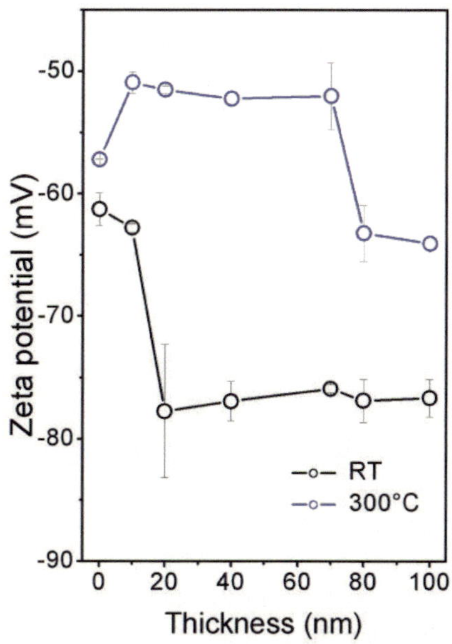

Figure 16. Dependence of zeta potential on the Au layer thickness for as-sputtered samples (RT) and the samples annealed at 300°C [63].

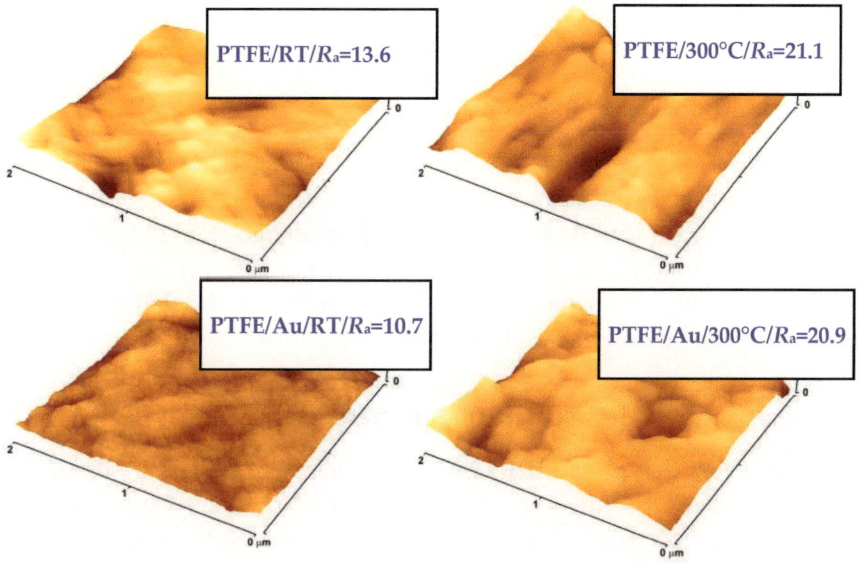

Figure 17. AFM images of pristine (PTFE) and Au coated (PTFE/Au) samples (thickness of 20 nm) before (RT) and after annealing at 300°C. Numbers in frames are measured surface roughnesses R_a in nm [63].

The rapid decrease in the sheet resistance occurs at the same layer thickness as the decrease in zeta potential. Both correlated changes are connected with creation of continuous, conductive gold coverage. Another interesting fact is that even for the layers with thicknesses above 80 nm, the values of the zeta potential measured on as-sputtered and annealed samples differ significantly. This can be due to higher fluorine concentration in the annealed samples and the fact that the C-F bond is more polar and exhibits higher wettability. It should be also noted that the value of the zeta potential may be affected by the surface roughness too. In general, it follows that the thicker the gold coverage the lower the zeta potential is, reflecting the electrokinetic potencial of metal itself.

The changes in the surface morphology after the annealing were studied by AFM. AFM scans of pristine and Au-coated (20 nm) samples before and after annealing are presented in Fig. 17. One can see that the annealing causes a dramatic increase in the surface roughness of the pristine polymer. Since the annealing temperature markedly exceeds PTFE glassy transformation temperature (T_g^{PTFE} = 126°C) the increase in the surface roughness is probably due to thermally induced changes of PTFE amorphous phase. The gold sputtering leads to a measurable reduction of the sample surface roughness. The reduction may be due to preferential gold growth in holes at the PTFE surface. Annealing of the gold-coated sample leads to significant increase of the surface roughness too. In this case, the increase is a result of both, the changes in the surface morphology of underlying PTFE and the changes in the morphology of the gold layer. After annealing, the surface roughness of pristine and gold-coated samples is practically the same. This finding is in contradiction with similar study accomplished on gold layers deposited on glass substrate [16]. Possible explanation of this fact probably lies in much better flatness of the glass substrate and in lower thermal stability of PTFE substrate during annealing.

4. Self-organized gold nanostructures

Purpose of this section lies in description of phenomena taking place during both interaction of polarized laser light with the surface of polymeric material and its subsequent coating by metal. It will be shown that modification of the polyethylenetherephtalate (PET) surface with linearly polarized light from pulsed KrF laser has a significant effect on the properties of subsequently deposited gold nanolayers and the choice of the deposition technique is crucial owing to the quality of prepared coatings.

4.1. Surface morphology and structure parameters

It has been shown [67] that by the KrF laser irradiation with several thousand of pulses a periodic ripple structure is formed at a PET surface for a fluence range from about 4.2 to 18.8 mJ·cm^{-2}. The ripples have a fluence-independent width Λ, which is given by the formula $\Lambda = \lambda/(1 - \sin(\theta))$, Eq. (1) [68], where λ is wavelength of a laser light used, n the effective refractive index of material, and θ the angle of incidence. Fig. 18 displays AFM images of pristine PET and PET irradiated at different laser fluences. The sample irradiated with the laser fluence of 3.4 mJ·cm^{-2} exhibits a rougher surface than the flat un-irradiated pristine PET. There is a noticeable modulation, although no ripple formation is visible. At higher laser fluences periodic ripple structures have developed in the irradiated area. At a laser fluence of 6.6 mJ·cm^{-2}, a regular and uniform coverage of the PET surface with ripples is reached. These results are based on AFM measurements, as those shown in Fig. 18. There is a good correlation between height and surface roughness of ripples over the whole laser fluence range shown in the figure. Both parameters reach the maximum value at a fluence of 6.6 mJ·cm^{-2}, which corresponds to a ripple height of about 90 nm.

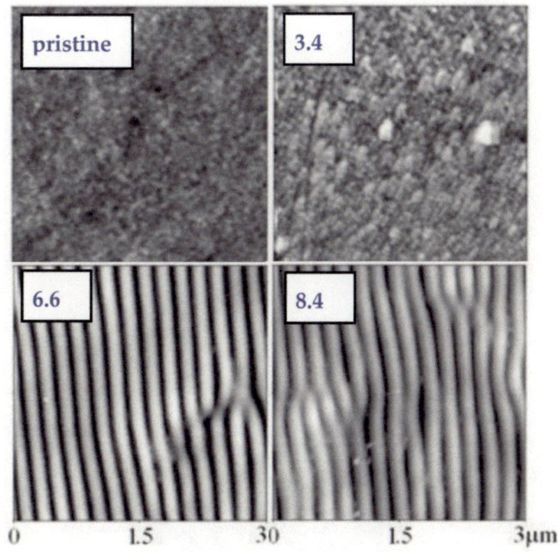

Figure 18. AFM images of the PET irradiated at different KrF laser fluences; the numbers in the inset refer to the laser fluence in mJ·cm^{-2} employed for irradiation of the PET foils, while pristine stands for unirradiated pristine PET [70].

The height and the roughness of the ripples as a function of the applied laser fluence are shown in Fig. 19.

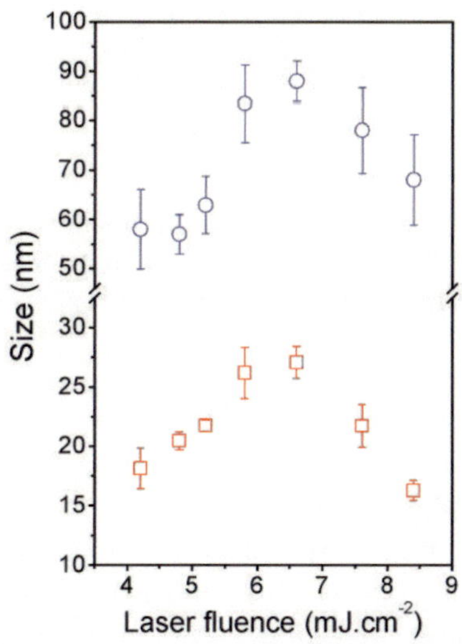

Figure 19. Dependence of the ripple height (○) and roughness (□) on the KrF laser fluence employed for the PET irradiation [70].

Fig. 20 shows AFM images of the PET irradiated under different incidence angles of the laser beam. For larger angles of incidence, the spacing between two neighboring ripples is wider. For the incidence angle of 0° and 22.5°, the observed spacing of the ripples is in good agreement with the value calculated by Eq. (1) with an effective index of refraction $n \approx 1.2$. The agreement for the incidence angle of 45° is less pronounced. The discrepancy may be due to changes of the polymer refractive index induced by the UV laser irradiation as reported earlier [69].

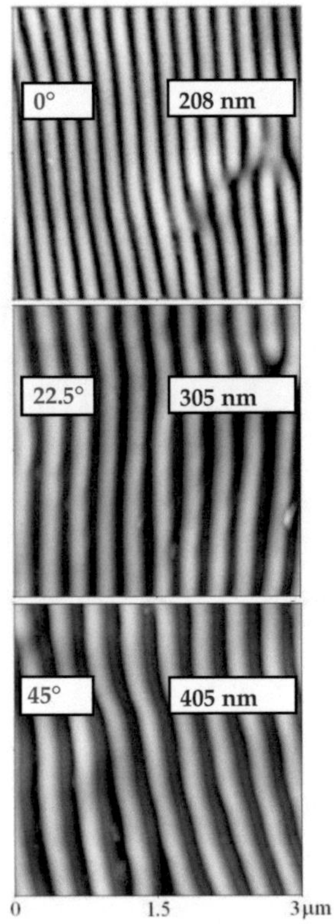

Figure 20. AFM images of the PET irradiated at a KrF laser fluence of 6.6 mJ·cm^{-2} under the different incidence angle of laser beam (0, 22.5 and 45°). The numbers in the insets in the upper left corner refer to the angle of incidence of the laser beam and in the insets in the upper right corner to the ripple period in nm [70].

FIB cuts of laser irradiated and gold coated PET samples were investigated by SEM (see Fig. 21). After sputtering, the gold is deposited in the form of "nanowires", which grow mainly at the ridges of the ripples. The FIB cuts reveal that there could be gaps between the individual wires and that the metal layer may be discontinuous. The width of the gold nanowires directly correlates to the width of the ripples formed before gold deposition. Additionally, a granularity is visible along the wires, but the FIB cut images suggest that the grains may be interconnected. The morphology of the gold layers deposited by evaporation is distinctly different. The gold is also deposited in the valleys of the ripple structure.

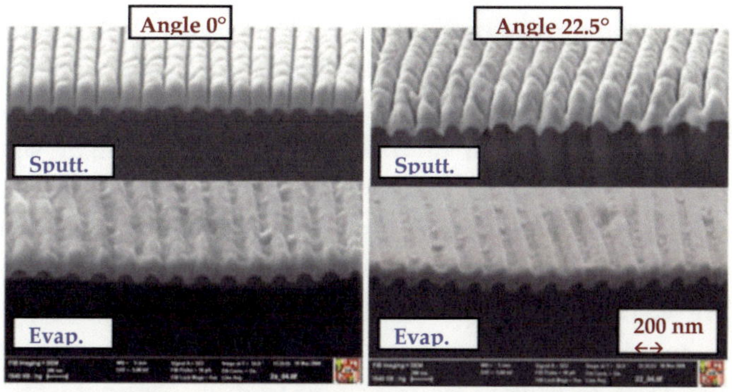

Figure 21. FIB-SEM images of the gold coatings on PET samples irradiated by a KrF laser under incidence angles 0 and 22.5° (fluence 6.6 mJ·cm^{-2}). The gold deposition was performed either by sputtering (Sputt.) and evaporation (Evap.) [70].

The nanowire structure of the sputtered gold layer can be observed also after gold deposition onto PET samples irradiated by the laser under an angle of incidence of 45°. Again, the evaporation of gold leads to a continuous coverage copying nanostructured polymer surface. The reason for the different observed gold morphologies after sputtering (nanowires) and evaporation (homogeneous gold coverage) is still unclear. The different particle energies in both processes are one possible reason. For sputtering, the particle energy may be considerably higher because of sample charging effect, while the evaporated materials should be slower (i.e., colder) and closer to thermodynamical equilibrium. The electrical charge of the sputtered particles can have also direct influence on layer formation, while the evaporated material should be mainly neutral. Other reasons may be the different deposition rates, which were a factor of two lower for sputtering than for evaporation, and possible differences of the substrate and gold layer temperature during the deposition in the two different techniques.

5. Summary

In summary, this chapter gives a comprehensive insight into the problematic of ultrathin gold films formed by physical deposition techniques on glass and polymeric substrates. Particular emphasis is given to the processes taking place during post-deposition annealing of prepared layers. In the case of glass substrate, the sputtering times and the layer effective thicknesses were chosen to span the region of the transition from discontinuous to continuous gold layer. For short sputtering times electrically discontinuous layers are obtained comprising discrete gold crystallites. The crystallite size in the as-sputtered samples is a monotonously increasing function of the sputtering time. The dependence of the lattice parameter of the gold crystallites forming the layer on the sputtering time is rather complicated with a rapid increase for shorter sputtering times and subsequent decrease for longer sputtering times. The decrease can be explained by relaxation processes in the thicker layers. The annealing has significant influence on the properties of the gold layers. For the annealed samples the lattice parameter practically does not depend on the sputtering time. The crystallite size first increases rapidly up to the sputtering time of 250 s, achieves a maximum and then decreases. The electrical properties (concentration, mobility of charge carriers and volume resistivity) and NIR optical properties of the gold layers change dramatically as the function of the sputtering time. Other significant changes, especially for electrically discontinuous layers, are observed as a result of the annealing. This is probably due to the different mechanism of free charge carrier transport, where also the quantum surface effects could be present in case of observed island structure mainly after annealing. For electrically continuous layer the concentration and the mobility are invariable. Similar behaviour exhibits also gold layers on polymeric substrate. From the measurement of the sheet resistance the transition from discontinuous to continuous gold coverage was found at the layer thicknesses of 10-15 nm for as-sputtered samples. After annealing at 300°C the transition point increases to about 70 nm, the increase indicating substantial rearrangement of the gold layer. The rearrangement is confirmed also by XPS measurement and an electrokinetic analysis. By XPS measurement contamination of the gold coated PTFE samples with carbon and the presence of oxidized structures, created during gold sputtering were proved. The annealing results in significant increase of the surface roughness of both pristine and gold sputtered PTFE.

Modification of the PET surface with linearly polarized light from pulsed KrF laser has a significant effect on the properties of subsequently deposited gold nanolayers and the choice of the deposition technique is crucial owing to the quality of prepared coatings. Subsequent deposition of 200 nm thick gold layer caused a decrease of the surface roughness. While by evaporation a continuous metal coverage is formed, copying nanostructured polymer surface, in the case of sputtering a nanowire-like structure of the gold coating can be observed. It was shown that the width of the nanowires can be tailored by the width of the ripples formed by preceding laser irradiation. We demonstrate a technique for the controlled patterning of polymer surfaces, including the creation of nanopatterned, regular gold structures (nanowires). In principle, this technique could be employed for the creation of metal-polymer composites with interesting electrical, mechanical, and optical properties, which could find novel applications in micro and nanotechnology.

Author details

Jakub Siegel, Ondřej Kvítek, Petr Slepička and Václav Švorčík
Institute of Chemical Technology Prague, Czech Republic

Zdeňka Kolská
University of J.E. Purkyne Usti nad Labem, Czech Republic

Acknowledgement

Financial support of this work from the GACR projects No. P108/11/P337, P108/10/1106 and 106/09/0125 is gratefully acknowledged.

6. References

[1] Rao CNR, Kulkarni GU, Thomas PJ, Edwards PP (2002) Size-Dependent Chemistry: Properties of Nanocrystals. Chem. eur. j. 8: 25-39.

[2] Roduner E (2006) Size Matters: Why Nanomaterials are Different. Chem. soc. rev. 35: 583-592.

[3] Daniel MC, Astruc D (2004) Gold Nanoparticles: Assembly, Supramolecular Chemistry, Quantum-Size-Related Properties, and Applications Toward Biology, Catalysis, and Nanotechnology. Chem. rev. 104: 293-346.

[4] Guoa S, Wang E (2007) Synthesis and Electrochemical Applications of Gold Nanoparticles. Anal. chim. acta 598: 181-192.

[5] Yonezawa T, Kunitake T (1999) Preparation of Anionic Mercapto Ligand-Stabilized Gold Nanoparticles and their Immobilization. Colloid surf. A 149: 193-199.

[6] Turkevich J, Stevenson PC, Hillier J (1951) A Study of the Nucleation and Growth Processes in the Synthesis of Colloidal Gold. Discuss faraday soc. 11: 55-59.

[7] Kim HJ, Jung SM, Kim BJ, Yoon TS, Kim YS, Lee HH (2010) Characterization of Charging Effect of Citrate-Capped AuNP Pentacene Device. J. ind. eng. chem. 16: 848-851.

[8] Xu SP, Zhao B, Xu WQ, Fan YG (2005) Preparation of Au-Ag Coreshell Nanoparticles and Application of Bimetallic Sandwich in SERS. Colloid surf. A 257-258: 313-317.

[9] Sánchez-López JC, Abad MD, Kolodziejczyk L, Guerrero E, Fernández A (2011) Surface-Modified Pd and Au NPs for Anti-Wear Applications. Tribol. int. 44: 720-726.

[10] Seino S, Kinoshita T, Otome Y, Maki T, Nakagawa T, Okitsu K, Mizukoshi Y, Nakayama T, Sekino T, Niihara K, Yamamoto TA (2004) Gamma-Ray Synthesis of Composite Nanoparticles of Noble Metals and Iron Oxides. Scripta mater. 51: 467-472.

[11] Kan CX, Zhu XG, Wang GH (2006) Single-Crystalline Gold Microplates: Synthesis, Characterization, and Thermal Stability. J. phys. chem. B 110: 4651-4656.

[12] Liu HB, Ascencio JA, Perez-Alvarez M, Yacaman MJ (2001) Melting Behavior of Nanometer Sized Gold Isomers. J. surf. sci. 491: 88-98.

[13] Siegel J, Lyutakov O, Rybka V, Kolská Z, Švorčík V (2011) Properties of Gold Nanostructures Sputtered on Glass. Nanoscale res. lett. 6: 96.

[14] Švorčík V, Siegel J, Šutta P, Mistrík J, Janíček P, Worsch P, Kolská Z (2011) Annealing of Gold Nanostructures Sputtered on Glass Substrate. Appl. phys. A 102: 605-610.

[15] Solliard C, Flueli M (1985) Surface Stress and Size Effect on the Lattice-Parameter in Small Particles of Gold and Platinium. Surf. sci. 156: 487-494.

[16] Švorčík V, Kvítek O, Lyutakov O, Siegel J, Kolská Z (2011) Annealing of Sputtered Gold Nano-Structures. Appl. phys. A 102: 747-751.

[17] Heo DN, Yang DH, Moon HJ, Lee JB, Bae MS, Lee SC, Lee WJ, Sun IC, Kwon IK (2012) Gold Nanoparticles Surface-Functionalized with Paclitaxel Drug and Biotin Receptor as Theranostic Agents for Cancer Therapy. Biomaterials 33: 856-866.

[18] Lee S, Chon H, Yoon SY, Lee EK, Chang SI, Lim DW, Choo J (2012) Fabrication of SERS-Fluorescence Dual Modal Nanoprobes and Application to Multiplex Cancer Cell Imaging. Nanoscale 4: 124-129.

[19] Choi KY, Liu G, Lee S, Chen XY (2012) Theranostic Nanoplatforms for Simultaneous Cancer Imaging and Therapy: Current Approaches and Future Perspectives. Nanoscale 4: 330-342.

[20] Curry AC, Crow M, Wax A. (2008) Molecular Imaging of Epidermal Growth Factor Receptor in Live Cells with Refractive Index Sensitivity Using Dark-Field Microspectroscopy and Immunotargeted Nanoparticles. J. biomed. opt. 13: 014022.

[21] Švorčík V, Kolská Z, Kvítek O, Siegel J, Řezníčková A, Řezanka P, Záruba K (2011) "Soft and Rigid" Dithiols and Au Nanoparticles Grafting on Plasma-Treated Polyethyleneterephthalate. Nanoscale res. lett. 6: 607.

[22] Švorčík V, Chaloupka A, Záruba K, Král V, Bláhová O, Macková A, Hnatowicz V (2009) Deposition of Gold Nano-Particles and Nano-Layers on Polyethylene Modified by Plasma Discharge and Chemical Treatment. Nucl. instrum. meth. B 267: 2484-2488.

[23] Wang ZL (2000) Characterizing the Structure and Properties of Individual Wire-Like Nanoentities. Adv. mater. 12: 1295-1298.

[24] Hu JT, Odom TW, Lieber ChM (1999) Chemistry and Physics in One Dimension: Synthesis and Properties of Nanowires and Nanotubes. Acc. chem. res. 32: 435-445.

[25] Hollensteiner S, Spiecker E, Dieker C, Jäger W, Adelung R, Kipp L, Skibowski M (2003) Self-Assembled Nanowire Formation During Cu Deposition on Atomically Flat Vse(2) Surfaces Studied by Microscopic Methods. Mater. sci. eng. C 23: 171-179.

[26] Milenkovic S, Nakayama T, Rohwerder M, Hassel AW (2008) Structural characterisation of gold nanowire arrays. J. cryst. growth 311: 194-199.

[27] Lide DR (1994) The Handbook of Chemistry and Physics 74th edn. Boca Raton: Chemical Rubber Company. 68 p.

[28] Greenwood NN (1984) Earnshaw A Chemistry of the Element. New York: Pergamon. 126 p.

[29] Kouklin NA, Kim WE, Lazareck AD, Xu JM (2005) Carbon Nanotube Probes for Single-Cell Experimentation and Assays. Appl. phys. lett. 87: 173901.

[30] Obataya I, Nakamura C, Han S, Nakamura N, Miyake J (2005) Nanoscale Operation of a Living Cell Using an AFM with a Nanoneedle. Nano lett. 5: 27-30.

[31] Gross GW, Wen WY, Lin JW (1985) Transparent I Electrode Patterns for Extracellular, Multisite Recording in Neuronal Cultures. J. neurosci. meth. 15: 243-252.

[32] Pine J (1980) Recording Action-Potentials from Cultured Neurons with Extracellular Micro-Circuit Electrodes. J. neurosci. meth. 2: 19-31.

[33] Švorčík V, Kasálková N, Slepička P, Záruba K, Bačáková L, Pařízek M, Lisa V, Ruml T, Macková A (2009) Cytocompatibility of Ar(+) Plasma Treated and Au Nanoparticle-Grafted PE. Nucl. instrum. meth. B 267: 1904-1910.

[34] Chithrani BD, Ghazani AA, Chan WCW (2006) Determining the Size and Shape Dependence of AuNP Uptake into Mammalian Cells. Nano lett. 6: 662-668.

[35] Rosi NL, Giljohann DA, Thaxton CS, Lytton-Jean AKR, Han MS, Mirkin CA (2006) Oligonucleotide-Modified Gold Nanoparticles for Intracellular Gene Regulation. Science 312: 1027-1030.

[36] Favier F, Walter EC, Zach MP, Benter T, Penner RM (2001) Hydrogen Sensors and Switches from Electrodeposited Pd Mesowire Arrays. Science 293: 2227-2231.

[37] Wan Q, Li QH, Chen YJ, Wang TH, He XL, Li JP, Lin CL (2004) Fabrication and Ethanol Sensing Characteristics of ZnO Nanowire Sensors. Appl. phys. lett. 84: 3654-3656.

[38] Yang F, Taggart DK, Penner RM (2009) Fast, Sensitive Hydrogen Gas Detection Using Single Palladium Nanowires That Resist Fracture. Nano lett. 9: 2177-2182.

[39] He B, Morrow TJ, Keating CD (2008) Nanowire Sensors for Multiplexed Detection of Biomolecules. Curr. opin. chem. biol. 12: 522-528.

[40] Cui Y, Wei QQ, Park HK, Lieber CM (2001) Nanowire Nanosensors for Highly Sensitive and Selective Detection of Biological and Chemical Species. Science 293: 1289-1292.

[41] Poborchii VV, Tada T, Kanayama T (2002) Photonic-Band-Gap Properties of Two-Dimensional Lattices of Si Nanopillars. J. appl. phys. 91: 3299-3305.

[42] Au FCK, Wong KW, Tang YH, Zhang YF, Bello I, Lee ST (1999) Electron Field Emission from Silicon Nanowires. Appl. phys. lett. 75: 1700-1702.

[43] Gotschy W, Vonmetz K, Lietner A, Aussenegg FR (1996) Thin Films by Regular Patterns of Metal Nanoparticles: Tailoring the Optical Properties by Nanodesign. Appl. phys. B 63: 381-384.

[44] Oates TWH, Keller A, Facsko S, Mücklich A (2007) Aligned Silver Nanoparticles on Rippled Si Templates with Anisotropic Plasmon Absorption. Plasmonics 2: 47-50.

[45] Busbee BD, Obare SO, Murphy CJ (2003) An Improved Synthesis of High-Aspect-Ratio Gold Nanorods. Adv. mater. 15: 414-416.

[46] Zhang XY, Zhang LD, Lei Y, Zhao LX, Mao YQ (2001) Fabrication and Characterization of Highly Ordered Au Nanowire Arrays. J. mater. chem. 11: 1732-1734.

[47] Mbindyo JKN, Mallouk TE, Mattzela JB, Kratochvilova I, Razavi B, Jackson TN, Mayer TS (2002) Template Synthesis of Metal Nanowires Containing Monolayer Molecular Junctions. J. am. chem. soc. 124: 4020-4026.

[48] Wirtz M, Martin CR (2003) Template-Fabricated Gold Nanowires and Nanotubes. Adv. mater. 15: 455-458.

[49] Ghanem MA, Bartlett PN, De Groot P, Zhokov A (2004) A Double Templated Electrodeposition Method for the Fabrication of Arrays of Metal Nanodots. Electrochem. commun. 6: 447-453.

[50] Švorčík V, Slepička P, Švorčíková J, Špírková M, Zehentner J, Hnatowicz V (2006) Characterization of Evaporated and Sputtered Thin Au Layers on Poly(Ethylene Terephtalate). J. appl. polym. sci. 99: 1698-1704.

[51] Kaune G, Ruderer MA, Metwalli E, Wang W, Couet S, Schlage K, Röhlsberger R, Roth SV, Müller-Buschbaum P (2009) In Situ GISAXS Study of Gold Film Growth on Conducting Polymer Films. Appl. mater. interf. 1: 353-362.

[52] Kolská Z, Švorčík V, Siegel J (2010) Size-Dependent Density of Gold Nano-Clusters and Nano-Layers Deposited on Solid Surface. Collect. czech. chem. C. 75: 517-525.

[53] Donor-Mor I, Barkay Z, Filip-Granit N, Vaskevich A, Rubinstein I (2004) Ultrathin gold island films on silanized glass. Morphology and optical properties. 16:3476-3483.

[54] Haupl K, Lang M, Wissmann P (1986) X-Ray-Diffraction Investigations on Ultra-Thin Gold-Films. Surf. interface anal. 9: 27-30.

[55] Shyjumon I, Gopinadhan M, Ivanova O, Quaas M, Wulff H, Helm CA, Hippler R (2006) Structural Deformation, Melting Point and Lattice Parameter Studies of Size Selected Silver Clusters. Eur. phys. j. D37: 409-415.

[56] Santos VL, Lee D, Seo J, Leon FL, Bustamante DA, Suzuki S, Majima Y, Mitrelias T, Ionescu A, Barnes CHW (2009) Crystallization and Morphology of Au/SiO(2) Thin Films Following Furnace and Flame Annealing. Surf. sci. 603: 2978-2975.

[57] Tauc J (1974) Amorphous and Liquid Semiconductors. Heidelberg: Springer. 202 p.

[58] Fischer W, Geiger H, Rudolf P, Wissmann P (1977) Structure Investigations on Single-Crystal Gold-Films. Appl. phys. 13: 245-253.

[59] Slepička P, Kolská Z, Náhlík J, HnatowiczV, Švorčík V (2009) Properties of Au Nanolayers on PET and PTFE. Surf. interface anal. 41: 741-745.

[60] Hovel M, Gompf B, Dressel M (2010) Dielectric Properties of Ultrathin Metal Films Around the Percolation Threshold. Phys. rev. B 81: 035402.

[61] Hodgman CD (1975) Handbook of Chemistry. Cleveland: Chemical Rubber. 264 p.

[62] Chopra K (1969) Thin Film Phenomena. New York: Wiley. 138 p.

[63] Siegel J, Krajcar R, Kolská Z, Hnatowicz V, Švorčík V (2011) Annealing of Gold Nanostrucutres Sputtered on Polytetrafluoroethylene. Nanoscale res. lett. 6:588.

[64] Švorčík V, Kotál V, Siegel J, Sajdl P, Macková A, Hnatowicz V (2007) Ablation and Water Etching of PE Modified by Ar Plasma. Polym. degrad. stabil. 92: 1645-1649.

[65] Siegel J, Řezníčková A, Chaloupka A, Slepička P, Švorčík V (2008) Ablation and water etching of plasma-treated polymers. Radiat. eff. deffect S 163: 779-788.

[66] Švorčík V, Řezníčková A, Kolská Z, Slepička P (2010) Variable Surface Properties of PTFE Foils. e-Polymers 133: 1-6.

[67] Siegel J, Slepička P, Heitz J, Kolská Z, Sajdl P, Švorčík V (2010) Gold Nano-Wires and Nano-Layers at Laser-Induced NanoRipples on PET. Appl. surf. sci. 256: 2205-2209.

[68] Bäuerle D (2000) Laser Processing and Chemistry. Berlin-Heidelberg-New York: Springer-Verlag. 378 p.

[69] Dunn DS, Ouderkirk AJ (1990) Chemical and Physical Properties of Laser-modified Polymers. Macrolmolecules 23:770-774.

[70] Siegel J, Heitz J, Švorčík V (2011) Self-organized Gold Nanostructures on Laser Patterned PET. Surf. coat. technol. 206:517-521.

Chapter 4

Low Temperature Wafer-Level Metal Thermo-Compression Bonding Technology for 3D Integration

Ji Fan and Chuan Seng Tan

Additional information is available at the end of the chapter

http://dx.doi.org/10.5772/48216

1. Introduction

1.1. Background of 3D integration

The past few decades have seen the rapid development in computing power and wireless communication achieved through invention of new technologies, discovery of new semiconductor materials and application of new fabrication processes. These innovations have enabled the placement of large number of high performance transistors which are commensurate with scaling on an integrated circuit and the transistor count doubles approximately every 18 months, which is famously known as Moore's law as described by Gorden Moore (Moore, 1998) and modified by Davide House (Kanellos, 2003). Since small and efficient system is always the ultimate objective for semiconductor industry development, 3D integration emerges as a suitable candidate for mainstream packaging and interconnection technology in the future as geometrical scaling is faced with unprecedented scaling barriers from the fundamental and economics fronts.

The development of vertically integrated devices could date back to the early1980s (Pavlidis&Friedman, 2009). One of the first successful 3D structure just includes a positive-channel metal oxide semiconductor (PMOS) and a negative-channel metal oxide semiconductor (NMOS) transistors which share the same gate electrode to reduce the total area of the inverter (Gibbons&Lee, 1980; Goeloe et al., 1981). After 30 years of continuous development, 3D integration technology has infiltrated into all the domains of semiconductor, such as combination of logic and memory circuit (Beyne, 2006), sensor packaging (Yun et al., 2005), heterogeneous integration of MEMS and CMOS (Lau et al., 2009), etc. More importantly, 3D integration technology is not only used for form-factor miniaturization, but also for achieving excellent performance, low power consumption, high integration capability, and low cost (Tan et al., 2011).

1.2. Motivation for research in 3D integration

In order to keep up with the needs of the astonishing development in the functionality of portable devices and of computational systems, integration technology has been investigated over the past three decades. 3D integration technology is widely defined as the technology which can integrate the disparate device layers in a multi-strata vertical stacking way (Young&Koester, 2010) with electrical interconnects by vertical inter-layer vias. Fig. 1 schematically presents the concept of 2D and 3D integration circuit. Further requirements on form factor reduction, performance improvement, and heterogeneous integration will make 3D integration a plausible choice as the next generation of microsystem manufacturing technology, as it can provide an excellent connection density higher than $10^4/mm^2$ (Beyne, 2006) for developing "More than Moore" scaling.

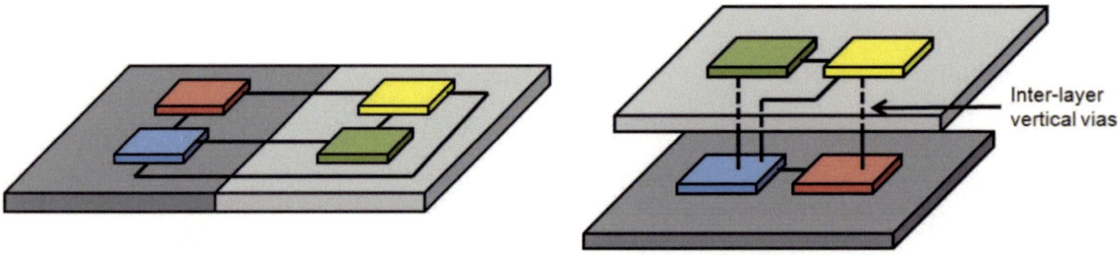

(a) 2D integration circuit (b) 3D integration circuit

Figure 1. Different approaches of integration technology.

The original purpose of 3D structure is system-size reduction. Traditional 2D integration technologies individually assemble different functional dies on a planar substrate or in a printed circuit broad. The packaging area of individual die is generally needed and an additional spacing between disparate functional blocks is typically required, thus reducing the integration density to a very low level. By stacking the device layers in a vertical way, a highly integrated circuit can be achieved. Since the substrate area is the first consideration, high integration density can increase the number of devices or functional blocks per chip area, which in turn miniaturizes the form-factor.

High performance requirement is another important reason for research in 3D integration. As the dimension of functional blocks continues to shrink and the emergence of large scale integration (LSI) technology, or even very large scale integration (VLSI) technology in recent years, the interconnects in an integrated circuit has begun to dominate the overall circuit performance. As a result of long interconnect length, interconnect latency and power consumption will increase. Therefore, the number of long wires is identified as the bottle-neck in the planar (2D) integration. In comparison with 2D design, 3D integration technology based on flip-chip, micro-bump connection and through silicon via (TSV) technologies can ease this interconnect bottle-neck and thus results in a lower propagation delay and power consumption. More importantly, in one synchronous operation mode, on-chip signal can only propagate in a limited distance. In other words, large chip size usually requires more clock cycles for on-chip signal to travel across the entire circuit. Using 3D stacking technology, more functional devices can be integrated in one synchronous region, thus increasing the computational speed.

The third, and maybe the most attractive, advantage of 3D integration technology is heterogeneous integration (Beyne, 2006). Although system-on-a-chip (SoC) is an attractive solution to integrate multiple functionalities on a single chip, specific optimization for each functional blocks on the same substrate may make SoC devices with large numbers of functional blocks very difficult to achieve. Furthermore, compatibility between different substrates might cause potential contamination or signal corruption. If high density 3D integration technology is available, it is a very attractive method for a "3D-SoC" device manufacturing. With this method, each functional block can be optimized separately and assembled in a vertical fashion. Since there is no common substrate, the problems caused by compatibility between different substrates are expected to be less severe. Fig. 2 shows an example for heterogeneous TSV-less integration method of CMOS and MEMS whereby the CMOS layer can be used as an 'active capping' layer for the sensitive MEMS layer. In order to provide a hermetic ambient for proper operation, the seal ring is formed in the trench of SOI MEMS wafer during the device layer DRIE etching. The electrode pad of MEMS is bonded to a connection pad on the CMOS die and will be routed to external by using lower metal layers in the CMOS chip. Metallization process can be realized during the SOI MEMS fabrication.

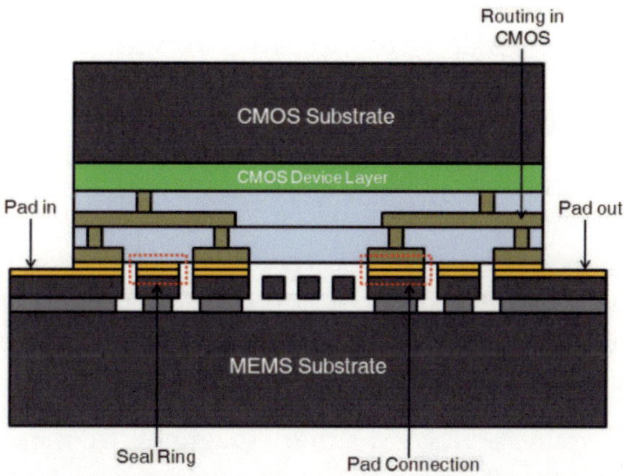

Figure 2. TSV-less 3D integration heterogeneous of MEMS and CMOS (Nadipalli et al., 2012).

2. Overview on bonding technologies in 3D integration

Bonding technologies have been reported as an imperative packaging and integration method for 3D IC stacking. It can be split into three schemes according to the fabrication approach: wafer to wafer, chip to wafer and chip to chip, as shown in Fig. 3. The ability of wafer to wafer bonding technology can effectively increase the throughput, making it a cost-effective manufacturing approach, but the unstable number of known-good-die (KGD) which is determined by the device layer might be the a drawback for this stacking method. Therefore, chip to wafer bonding and chip to chip bonding can assure that the vertical stacking will be only executed with the good dies. Since mass production is the primary commercial and manufacturing consideration in future, both chip-to-wafer and wafer-to-wafer technologies will gradually become the mainstream for 3D stacking and packaging (Ko&Chen, 2010).

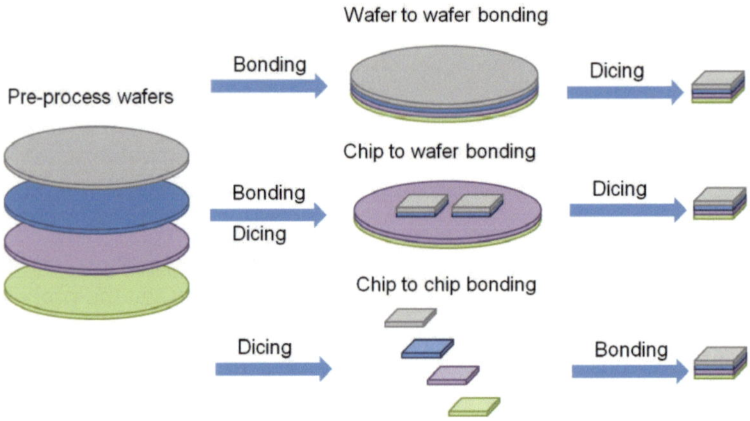

Figure 3. Different bonding technologies for 3D Integration circuit according to fabrication approach.

Based on the bonding materials, bonding technology can fall into dielectric bonding and metallic bonding. Since the dielectric materials are used, the device layers are isolated from each other and a "via-last" process is followed. Devices layers are firstly bonded in a vertical stack, and then the vertical vias are etched through the devices layers for vertical interconnects between each layer. Therefore, high aspect ratio vias are usually needed. The most leading dielectric bonding methods used in 3D integration include adhesive bonding and oxide fusion bonding.

Adhesive bonding, also known as polymer bonding, usually uses polymers and inorganic adhesive as the intermediate bonding materials. Since a layer of polymer or inorganic adhesive is always spun before bonding, it is very suitable for non-uniform surfaces and for bonding at low temperature. Benzocyclobutene (BCB) and SU-8 are the most common materials used in 3D integration, since high bonding strength can be easily achieved with these two polymer materials (Niklaus et al., 2001; Pan et al., 2002).

Oxide fusion bonding requires a very low surface roughness (root mean square roughness < 1 nm) and the process is often followed by a post-bonding annealing. The bonding step for fusion bonding refers to spontaneous adhesion process of two wafers when they are placed in direct contact. The surface activation which enables the wafer pair to have a stronger spontaneous adhesion is usually applied before bonding. This bonding technology is not only limited between Si-to-Si and SiO_2-to-SiO_2, but some high-*k* dielectric materials, such as Al_2O_3, HfO_2, and TiO_2 (Chong&Tan, 2009) are also employed to achieve a higher bonding strength for a given anneal temperature and duration.

Device layers bonded with a conductive metallic layers is a very attractive choice, as it allows "via-first" and "via-middle" approaches for 3D IC integration. Therefore, the requirement for high aspect ratio via can be relaxed. On the other hand, metal is a good heat conductor which will help to circumvent the heat dissipation problem encountered in 3D ICs. At the same time, the use of metal as bonding material in 3D applications allows the electrical contact and mechanical support to be formed between two wafers in one simultaneous step. Examples of such bonding technology include metal diffusion bonding and eutectic bonding will be presented in details in next section.

Dielectric bonding and metallic bonding can be combined to one emerging approach for 3D integration as well. The research work by McMahon *et al.* (Mcmahon et al., 2005) presents a wafer bonding of metal/adhesive damascene-patterned providing inter layer electrical interconnects via Cu-Cu bonding and mechanical support via adhesive bonding (BCB) of two wafers in one unit processing step. IMEC (Interuniversity Microelectronics Centre, Belgium) developed this technology and was formally named as "Hybrid Bonding" (Jourdain et al., 2007; Jourdain et al., 2009).

3. Low temperature wafer-level metal thermo-compression bonding for 3D integration

Pure metal and alloy material are widely used in bonding technology for 3D integration. The description in this section is specific on two types of metal based low temperature thermo-compression bonding technologies: copper diffusion bonding and copper/tin eutectic bonding. The following description includes the comparison of different metal bonding materials, the principle of bonding process and performance of reported work. Oxide fusion bonding which is also widely investigated in 3D integration is included for comparison at the end of this section as well.

3.1. Why low temperature?

As the name implies, thermo-compression bonding contains two important elements: heat and pressure. Metal bonding surfaces are brought into contact with the application of force and heat simultaneously. Atomic motion (for metal diffusion bonding) or alloy formation (for eutectic bonding) will occur at the bonding interface during this process. Due to surface oxidation and some specific alloy formation, high temperature is usually required for achieving high bonding quality, but practical packaging and integration should be achieved at adequately low temperature (typically 300°C or below) for prefabricated devices which are sensitive to high temperature processing owing to thermal budget limitation, the post-bonding thermo-mechanical stress control, and alignment accuracy improvement (Tan et al., 2009). Although it is commonly known that the quality of thermo-compression bonding can usually be ameliorated when the bonding temperature increases, the current mainstream research focuses on achieving high bonding quality at temperature as low as possible for the considerations of cost reduction and high throughput manufacturing.

3.2. Metal diffusion bonding

Metal diffusion bonding is also referred as pressure joining, thermo-compression welding or solid-state welding. Bonding interfaces will fuse together due to atomic interaction under heat and pressure.

3.2.1. Comparison of different diffusion bonding materials

The common metal materials for metal diffusion bonding are aluminum (Al-Al), gold (Au-Au) and copper (Cu-Cu). Table 1 shows a comparison of physical properties of these metals in the context of metal diffusion bonding. Among these metals, Al-Al bonding is hard to achieve with low bonding temperature and low bonding force, most likely because it gets oxidized readily in ambient conditions. In addition, its relatively higher coefficient of thermal expansion (CTE) in comparison with that of silicon wafer will result in larger wafer bow during cooling. This poses difficulty in achieving high quality bonding for Al-Al especially across large area. On the other hand, even though the bonding temperature for Au-Au is generally about 300°C, its prohibitively high cost is the major roadblock for widespread use except for high end applications. Cu emerges as an attractive choice in terms of its lower cost and the ability to bond Cu at moderately low temperature. Furthermore, Cu presents a number of advantages in terms of its physical properties that suit the final application such as better electrical conductivity, mechanical strength and electro-migration resistance. Therefore, with these superior material properties, low temperature Cu bonding will be the candidate for mainstream 3D integration application.

Material	Al-Al	Au-Au	Cu-Cu
Temperature (°C)	450	~300	250/300
Resistivity (μΩcm)	2.67	2.20	1.69
Melting Point (°C)	660.4	1064.4	1083
CTE (×10^{-6} /K)	23.5	14.1	17.0
Thermal Conductivity (W/ m K)	237	318	401
Cost (US$/LB)	1.1798	25881.44	4.4198

Table 1. Comparison of different metal diffusion bonding technologies (Fan et al., 2011).

3.2.2. Fundamentals of low temperature copper diffusion bonding

Fig. 4 illustrates the formation principle of low-temperature Cu diffusion bonding. In order to isolate the substrate from the Cu bonding film, a thin film of dielectric, such as SiO_2 is firstly deposited as the precursor (Fig. 4. a). Subsequently, the barrier layer, such as Ti or Ta which is used to thwart excessive Cu diffusion into Si and to ameliorate the adhesion between substrate and Cu film, and a thin Cu seed layer are deposited (Fig. 4. b). After that, Cu electroplating is applied to the required thickness of Cu layer depending on the applications (usually from several to a dozen of μm), followed by chemical mechanical planarization (CMP) (Fig. 4. c). Finally, the wafer pair is brought into contact in nitrogen (N_2) or vacuum ambient under a contact force (e.g. 2500 mbar for a 6 inch wafer) and held typically at 300°C (or 250°C) for some time (e.g. 30 min and above) (Fig. 4. d).

Since the bonding temperature is fixed at low level, key parameters of low temperature Cu diffusion bonding are bonding duration and bonding force. During bonding, Cu atoms acquire sufficient energy to diffuse rapidly and Cu grain begins to grow. In order to obtain a higher bonding strength, Cu diffusion must happen across the bonding interface and the grain growth also needs to progress across the interface. If the bonding duration is insufficient (e.g. 10 min or below), the inter-diffusion of Cu atoms across the bonding interface is limited. Thus Cu grain formation stops at bonding interface and bonding strength is reduced. More importantly, Cu gain across the bonding interface will reduce the number of grain boundaries, which will provide a high conductivity at the bonding interface. However, this difficulty can be overcome by an anneal step after bonding. Fig. 5 presents the comparison of bonding strength, measured by die shear strength testing, without and with anneal step after a short bonding duration. The samples are bonded and annealed at 250°C for 15 min and 1hr, respectively. The bonding strength presents a significant improvement when short bonding duration is followed by a anneal process.

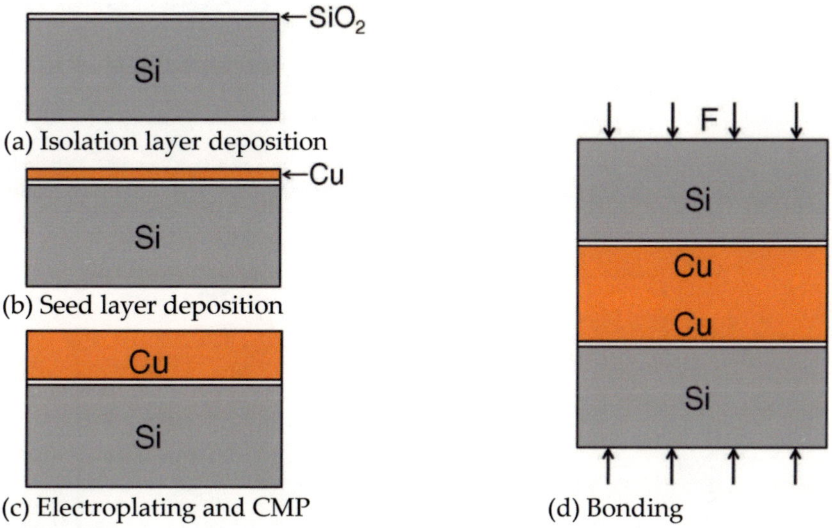

Figure 4. Schematic showing the principal of formation of Cu diffusion bonding.

Cu diffusion bonding is based on Cu atom migration and grain growth. Therefore, the wafer pairs must be brought into an intimate contact at the atomic level by a uniform bonding force. Wafer bow and surface contamination are the critical factors that affect the bonding uniformity. The surface contamination can usually be reduced with tighter particle control or some surface treatment before bonding which will be presented in next section. As shown in Fig. 6, a Cu-coated wafer exhibits a wafer bow of ~15.9 μm based on wafer curvature measurement using a laser beam. This is a direct result of the huge difference in the CTE between Cu and Si (17 and 3 $\times 10^{-6}$ /K, respectively), and it might become bigger during the bonding process. Appropriate bonding force enables us to eliminate the drawback brought by the wafer bow and perform a highly uniform bonding. For the wafer pairs that exhibit wafer bow lower than 20 μm, high bonding uniformity can usually be achieved under a contact pressure of ~2000 mbar at 300°C for 1 hour without post-bonding anneal.

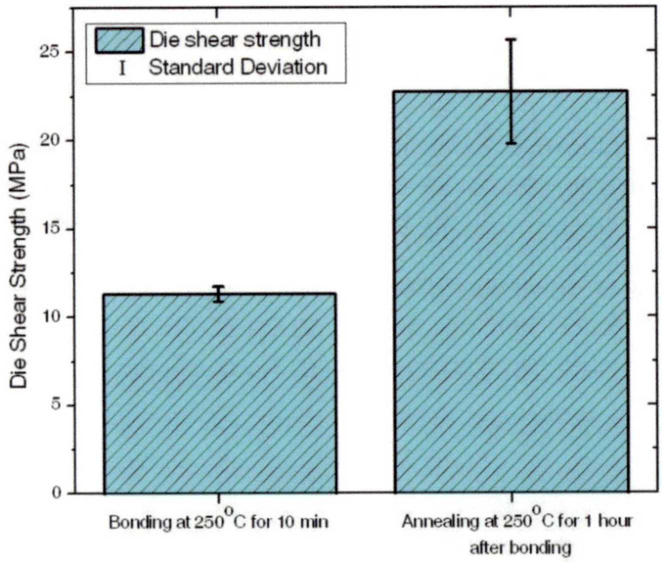

Figure 5. Comparison of bonding strength, without and with an anneal step after a short bonding duration.

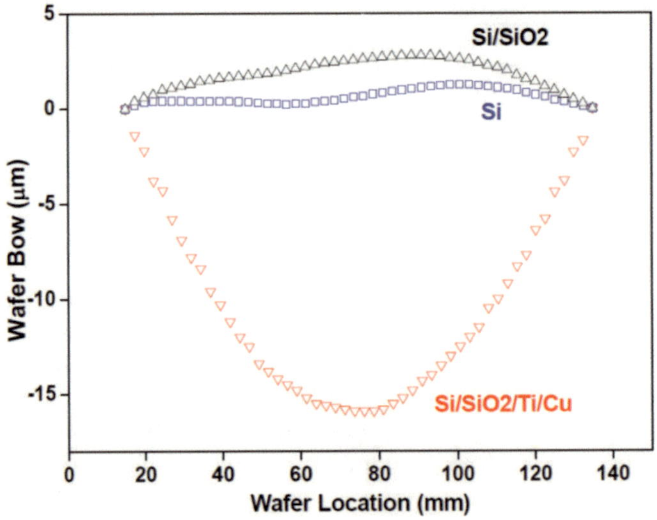

Figure 6. Pre-bonding wafer bow based on wafer curvature measurement (Tan et al., 2011).

3.2.3. Surface treatment before copper diffusion bonding

The surface condition often refers to the oxidation at the Cu bonding surface. Since Cu surface oxidizes readily in ambient air to form cuprous oxide (red oxide) and cupric oxide (black oxide). These oxide layers impose a barrier to successful diffusion bonding at low temperature.

3.2.3.1. Wet etch method

In order to remove the surface oxide, surface treatments by soaking the wafers in acetic acid or dilute hydrochloric acid and followed by a forming gas purge in the bonding chamber are usually applied immediately before bonding (Tadepalli&Thompson, 2003). This removal process can be described by the following chemical equations (1) and (2):

$$2H^{+} + CuO \rightarrow Cu + 2H_2O \quad (1)$$

$$2H^{+} + Cu_2O \rightarrow 2Cu + 2H_2O \quad (2)$$

Since this reaction takes place very rapidly, the immersion is completed in a few minutes. The research work by Jang *et al.* (Jang et al., 2009) indicates that, for the consideration of bonding strength, the immersion time must less than 5 min for a thickness of bonding layer around 500 nm. If long time immersion is applied, the bonding strength will be reduced due to the decrease in plastic dissipation energy near the interfacial crack tips with thinner Cu film thickness caused by over etching.

3.2.3.2. Forming gas anneal

Oxygen content in the bonding layer can be reduced by pre-bonding forming gas anneal. Forming gas is a mixture of hydrogen and nitrogen (typically 5%H_2:95%N_2, by volume). The reactions with the Cu oxides are exothermic and can be principally represented as follows in (3) and (4):

$$H_2 + CuO \rightarrow Cu + 2H_2O \quad (3)$$

$$H_2 + Cu_2O \rightarrow 2Cu + 2H_2O \quad (4)$$

This pre-bonding anneal is an *in-situ* clean process, which presents no re-oxidation risk before bonding. Compared with anneal at high temperature, long anneal duration (e.g. 1 hour) at low temperature (typically at 250 or 300 °C) is preferred to eliminate the oxygen in the bonding layer, as high temperature takes the potential risk of unwanted damage in the device layer.

3.2.3.3. Self-assembled monolayer (SAM) passivation

Even though the surface oxide removal by wet cleaning and oxide content reduction in the bonding layer by forming gas anneal have been widely investigated with some success, surface contamination of particles can still remain a challenge. Recently, a novel surface treatment using self-assembled monolayer (SAM) of alkane-thiol to passivate clean Cu surface immediately after metallization is applied. SAM application was first applied in wire bonding by IMEC in the area of microelectronic manufacturing (Whelan et al., 2003). Subsequently, this method is applied in the domain of fluxless solder, flip-chip bonding, and wafer-level Cu diffusion bonding. SAM of alkane-thiol formed by linear alkane-thiol molecules (CH_3-$(CH_2)_{n-1}$-SH, n = number of carbon), and it can be dissolved in ethanol to a concentration of 1 mM for the passivation application in wafer-level Cu diffusion bonding.

The process flow includes post-metallization adsorption and pre-bonding *in-situ* desorption to provide clean Cu surfaces for bonding.

Fig. 7 shows schematic of the process flow used in low temperature Cu diffusion bonding with SAM application. Wafers are immersed immediately into the solution of alkane-thiol after Cu metallization. Due to its specific high affinity functional head group (thiol, S-H) towards Cu surface, alkane-thiol can readily adsorb onto the Cu surface and rearranged into a uniform organic monolayer. This SAM layer provides temporary protection to the Cu surface. Subsequently, the SAM layer will be desorbed effectively with an annealing step in inert N_2 ambient to recover the clean Cu surface for the final bonding at low temperature. Research work in Nanyang Technological University (NTU) indicates that anneal for 30 min at 250 °C can efficiently desorb the SAM layer formed after 3 hr of immersion time in the solution.

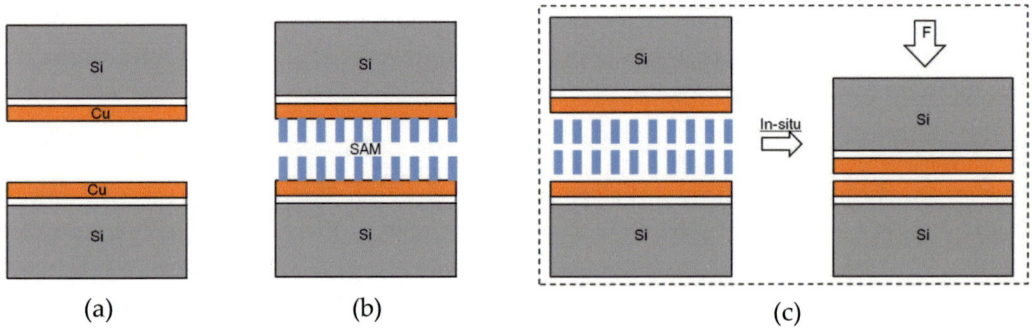

Figure 7. The application of self-assembled monolayer (SAM) as a passivation layer on Cu surface for bonding enhancement at lower temperature. (a) A pair of Si wafers with Cu metallization; (b) Immersion in alkane-thiol solution: SAM absorption; (c) Pre-bonding SAM desorption by thermal means and boding (Tan et al., 2012).

Fig. 8 shows the cross-section TEM images of the bonded Cu layers. The micrographs clearly confirm the success of Cu–Cu bonding in both samples. Fig. 8(a) is taken from bonded sample without SAM treatment. There is limited grain growth across the bonding interface and the original bonding interface is clearly seen (marked with arrows). In Fig. 8(b) which is taken from bonded sample with SAM treatment, the original bonding interface has disappeared. Cu grains extend across the bonding interface and a wiggling grain boundary is observed (marked with arrows).As can be seen, one Cu grain even extends the entire bonded Cu layer thickness sandwiched by the Ti capping layers (marked with white dotted line).

3.2.4. Performance of state-of-the-art copper diffusion bonding

Low temperature Cu diffusion bonding is gradually becoming the mainstream bonding technology for 3D integration as it allows the formation of electrical contact, mechanical support, and hermetic seal in one simultaneous step. Therefore, these three parameters are usually presented as the key performance matrix for metal based bonding quality.

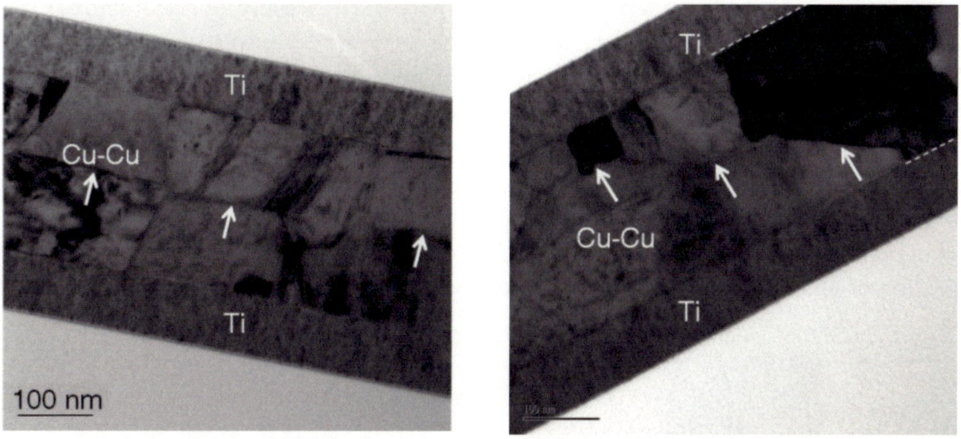

(a) Without SAM passivation (b) With SAM passivation and desorption

Figure 8. TEM micrographs of bonded Cu layers (Tan et al., 2009).

3.2.4.1. Electrical characterization

Early study of contact resistance of bonded Cu interconnects is presented by Chen *et al.* (Chen et al., 2004) in MIT (Massachusetts Institute of Technology, USA). The measurement results using Kelvin structure indicate that a specific contact resistance of bonding interfaces of approximately $10^{-8}\Omega.cm^2$ is obtained. A resent research by Peng *et al.* (Peng et al., 2011) demonstrate an excellent specific contact resistances of bonding interface using SAM as the surface treatment of about $2.59\times10^{-9}\Omega.cm^2$. This work has also demonstrated a daisy chain of at least 44,000 contacts at 15μm pitch connected successfully, and the misalignment of ~ 2 μm (Fig. 9).

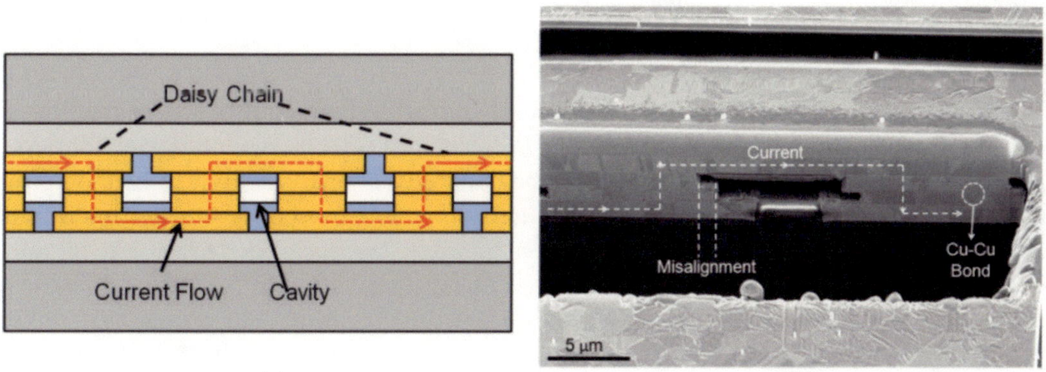

(a) (b)

Figure 9. (a) Face-to-face daisy chain stacking architecture of two wafers using Cu-Cu bonding. (b) Cross sectional FIB image of the 15 μm pitch Cu-Cu bonding (Peng et al., 2011).

No open failure is detected during measurement up to 44,000 nodes, as shown in Fig. 10. a. The sample with 10,000 bonding nodes is subjected to temperature cycling test (TCT) with temperature ranging from -40 °C to 125 °C. It is observed that the electrical continuity is maintained even after 1,000 thermal cycles (Fig. 10. b). In freshly bonded sample (before TCT test), the resistance of the daisy chain is estimated from *I-V* plot and each node (consists of Cu lines and contact) is estimated to have ~26.1 mΩ of resistance, and a slight increase of the node resistance up to ~29 mΩ at 1,000 temperature cycles. This slight increase is due to oxidation of the exposed Cu lines as a result of complete removal of the top wafer after bonding (since there is no TSV). The results suggest that the robustness of the Cu-Cu bond is maintained. This high connection density of up to 4.4 × 10^5/cm^2 and its reliability provides a feasible platform of high IC-to-IC connection density suitable for future wafer level 3D integration of IC to augment Moore's Law scaling.

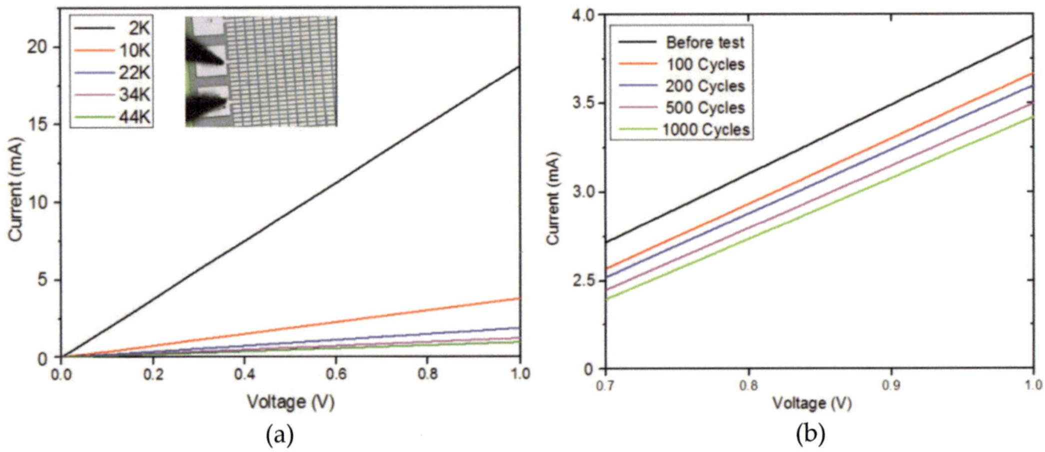

Figure 10. (a) *I-V* characteristic of the Daisy chain measured from 2,000 to 44,000 contacts (each interval = 2,000 contacts). The contacts are connected continuously and ohmic behavior is exhibited; (b) *I-V* characteristics of daisy chain before and after temperature cycling test (Peng et al., 2011).

3.2.4.2. Mechanical test

Besides the die shear strength test, four-point bending method is also widely employed for strength of mechanical support analysis (Huang et al., 2005). The interfacial adhesion energy between two bonded thin films can be qualitatively analyzed by this method. The earlier work by Tadepalli *et al.* (Tadepalli&Thompson, 2003) presents a superior interfacial adhesion energy of 11 J/m^2 at Cu diffusion bonding interface bonded at 300 °C and indicates that this value is superior than that of industry-standard SOI wafer. A recent work by Kim *et al.* (Kim et al., 2010) shows a short time bonding with post-anneal at 300 °C for 1 hour can also get a high interfacial adhesion energy around 12 J/m^2 which is much higher than critical bonding strength required (>5 J/m^2) by the subsequent processes such as grinding. A summary of interfacial adhesion energy achieved for wafer pairs with and without SAM passivation bonded at 250 °C for 1 hr is shown in Fig. 11. The average interfacial adhesion energy obtained with and without SAM passivation goes up to 18 J/m^2 and 12 J/m^2, respectively. Compared with the results from other literature, this interfacial adhesion energy obtained at low temperature is comparable or even better. The daisy chain bonding presented earlier exhibits high bonding strength as well, since bonded Cu structures need to provide sufficiently mechanical strength to sustain the shear force during wafer thinning.

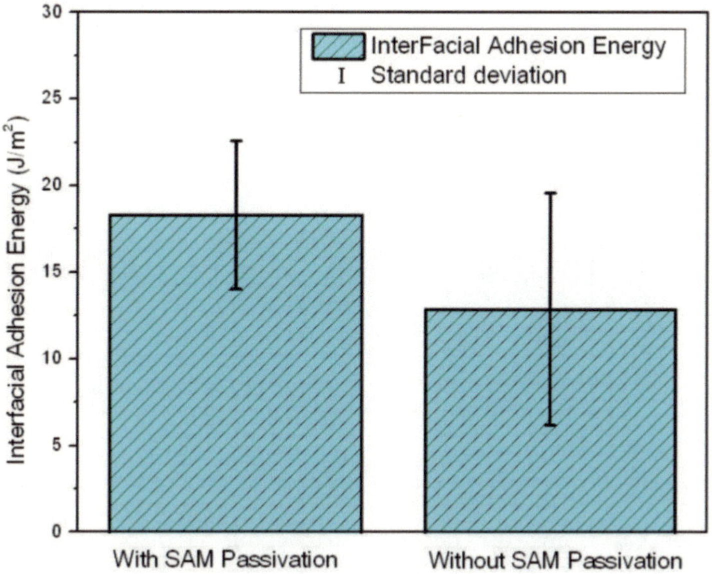

Figure 11. Interfacial adhesion energy for samples bonded at 250 °C.

3.2.4.3. Hermeticity detection

In an integrated 3D microsystems, micro- and nano-scale devices such as micro-electro-mechanical system (MEMS), microelectronic devices and optoelectronic devices, a hermetic ambient is commonly needed for proper operation with very low or without oxygen and water vapor content. The objective of hermetic packaging is to protect these devices against harsh environmental corrosion and potential damage during processing, handling and operation. Hermetic encapsulation can be also achieved by metal diffusion bonding. Hermeticity test, which consists of over-pressure storage in a helium bomb chamber and leak rate measurement with a mass spectrometer, is based on specifications defined in the MIL-STD, a standard commonly applied for microelectronics packaging. Hermetic packaging by Au diffusion bonding at 400°C demonstrated by Xu *et al.* (Xu et al., 2010) achieve a helium leak rate on order of 10^{-9} atm.cm^3/sec based on the MIL-STD-883E method 1014.9 specification. A research of Al diffusion bonding at 450°C by Yun *et al.* (Yun et al., 2008) presents a excellent result of helium leak rate of the order of 10^{-12} atm.cm^3/sec based on the MIL-STD-750E method 1071.8 specification.

The research work focusing on hermetic encapsulation with Cu diffusion bonding at low temperature in NTU exhibits outstanding helium leak rate based on the MIL-STD-883E method 1014.9 specification. Fig. 12 shows an average helium leak rate and standard deviation for cavities with the seal ring size of 50 μm sealed by Cu diffusion bonding at 250 °C and 300 °C respectively with proper surface preparation and control. These values are at least one order of magnitude smaller than the reject limit (5×10^{-8} atm.cm^3/sec) defined by the

MIL-STD-883E standard and is very attractive for packaging of devices that require high level of hermeticity and for heterogeneous integration of different micro-devices.

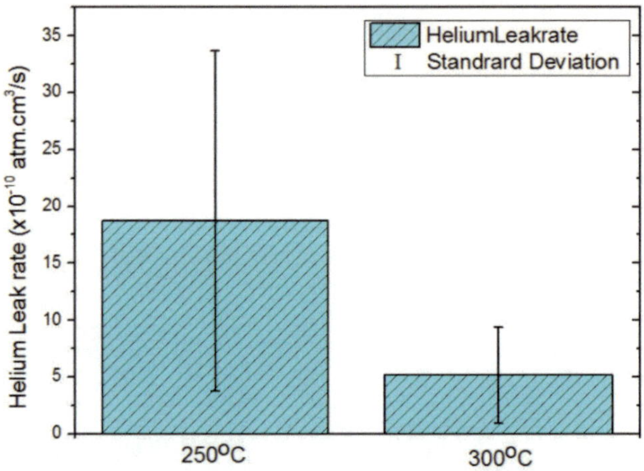

Figure 12. Average helium leak rate and standard deviation for cavities sealed at low temperature.

The reliability of Cu frame for hermetic packaging is also investigated through a temperature cycling test (TCT) from -40 to 125 °C up to 1000 cycles and a humidity test based on IPC/JEDEC J-STD-020 standard: (1) Level 1: 85°C/85%RH, 168hr; (2) Level 2: 85°C/ 60%RH, 168hr; and (3) Level 3: 30°C/ 60%RH, 192hr. The humidity test is applied from level 3 to level 1 in an ascending order in terms of rigor. In addition, an immersion in acid/base solution is applied to verify the corrosion resistance of the Cu frame for hermetic application. Table 2 shows some detected helium leak rate of sealed cavities with the seal ring size of 50 μm. Excellent reliability results of Cu-to-Cu wafer-level diffusion bonding at low temperature are maintained after a long term temperature cycling test with extreme low/high temperature swing, prolonged storage in humid environment, and immersion in acid/base solution.

TCT test	Before test	7.7×10^{-10} atm.cm^3/sec
	After 500 cycles	7.9×10^{-10} atm.cm^3/sec
	After 1000 cycles	7.9×10^{-10} atm.cm^3/sec
Humidity test	Before test	6.0×10^{-10} atm.cm^3/sec
	Level 3	7.5×10^{-10} atm.cm^3/sec
	Level 2	7.0×10^{-10} atm.cm^3/sec
	Level 1	8.0×10^{-10} atm.cm^3/sec
Corrosion test	Before test	7.1×10^{-10} atm.cm^3/sec
	Acid corrosion	5.9×10^{-10} atm.cm^3/sec
	Base corrosion	5.4×10^{-10} atm.cm^3/sec

Table 2. Detected helium leak rate after TCT test, humidity test, and corrosion test.

3.3. Eutectic bonding

Eutectic bonding is another metal based bonding technology for advanced MEMS packaging and for 3D integration. This technology, which is also referred as eutectic soldering and solid-liquid inter-diffusion bonding, stacks two wafers by intermediate eutectic compounds formation. The bonding interfaces will be fused together due to intermetallic phase formation. An important feature of eutectic bonding is the melting of intermediate eutectic metals and formation of the alloys that facilitate surface planarization and provide a tolerance of surface topography and particles.

3.3.1. Different alloy for eutectic bonding

The intermediate eutectic bonding layer is usually composed of a binary (or more) metal system. One with high melting point noble metal (like gold, silver and copper) and the other one with low melting point metal (like tin and indium) are used as intermediate eutectic metals which form intermetallic compounds during bonding. At present, the commonly used materials include Cu/Sn, Au/Sn, Au/Si, and Sn/Pb (Ko&Chen, 2010). Table 3 shows some eutectic metal system bonding temperature and their melting point. Since the eutectic point of two metals is lower than their melting points, the eutectic bonding can be usually achieved at low temperature. For example, in Cu/Sn, the bonding temperature is 150-280 °C. However, the temperature needed is still too high for some applications (e.g. Au/Si: 380 °C). The bonding temperature for Sn/Pb is only 183 °C, but this approach is not suitable for all electronic products due to the lead-free requirement.

Material	Bonding temperature	Material	Melting Point
Cu/Sn	150-280 °C	Cu	660.4 °C
Au/Sn	280 °C	Au	1064.4 °C
Au/Si	380 °C	Sn	213.9 °C
Sn/Pb	183 °C	Pb	327.5 °C

Table 3. Commonly used eutectic alloys bonding temperature and melting point of each metal

3.3.2. Fundamental of eutectic bonding

In following sections, the basic principles of eutectic bonding based on the binary metal systems Cu/Sn which is one of the best-investigated and well-established metal systems will be presented in detail.

3.3.2.1. Intermetallic compound formation

The bonding process relies on intermetallic compounds formed by inter-diffusion of the intermediate eutectic metal layers when they are brought into intimate contact at the specific bonding temperature. The first intermetallic compound formed between Cu and Sn is the metastable η-phase Cu_6Sn_5, and then the Cu_3Sn ε-phase starts to form at Cu to Cu_6Sn_5

interface. Fig. 13 presents a typical intermetallic compound formation during bonding. This process is terminated when all Sn is consumed to form Cu_3Sn, since the binary metal systems is thermodynamically stable while no non-reacted Sn remains (Munding et al., 2008). If the bonding time is insufficient, the transformation fails to complete. The joint presents a potential risk of resistance in high temperature environment, as the melting point of Cu_6Sn_5 is 415°C, while that of Cu_3Sn is 676°C.

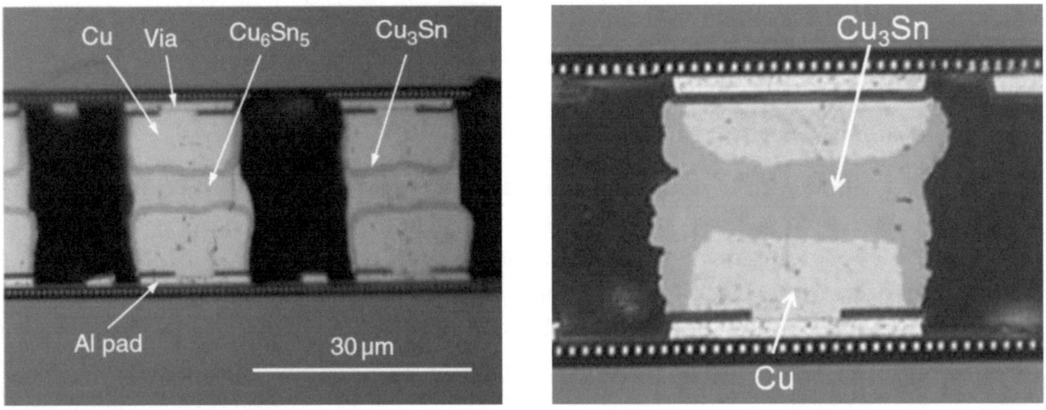

(a) (b)

Figure 13. (a) Cross sectional view of typical intermetallic compounds formation, from η-phase Cu_6Sn_5 to ε-phase Cu_3Sn; (b) Cross sectional view of completely alloyed joint, only ε-phase Cu_3Sn (Munding et al., 2008) [Copyright of Springer].

3.3.2.2. Temperature profile

A typical temperature profile for Cu/Sn eutectic bonding is shown in Fig. 14. Two bonding systems are widely used: Cu/Sn-Cu bonding and Cu/Sn-Sn/Cu bonding (Fig. 15). For Cu/Sn-Cu bonding system, the temperature ramping rate higher than 6 °C/s (Munding et al., 2008) is preferred, since the fast ramping rate after contact is beneficial to preserve most of reactive Sn. If Sn would have reacted with Cu during this period in the Cu/Sn stack, the Sn may be insufficient at the bonding interface for subsequent Cu/Sn-Cu diffusion. In this case, the delay between Sn melting and molten Sn wetting Cu surface is the other key parameter for metal system design and determines the necessary amount of Sn in the bonding process. In general, the Cu and Sn thickness should be related as $d_{Cu} \geq 1.5 d_{Sn}$. On the contrary, for Cu/Sn-Cu/Sn bonding, it is believed that a slow ramping rate is beneficial for reducing the flow of any excess Sn. More Sn would have reacted with Cu when temperature increases, and thus less pure liquid Sn is available at the bonding interface for combination (Lapadatu et al., 2010).

3.3.3. Performance of state-of-the-art eutectic bonding

The intermetallic compound formation is diffusion controlled which is directly related to the temperature. Below the melting temperature of Sn, the reaction is slow, but when the Sn begins to melt the reaction speed can be accelerated to an extremely high level. In order to control the diffusion of soldering process and to prevent solder consumption before bonding, a thin buffer layer can be deposited between Cu and Sn. With a thin buffer layer, the bonding process begins with the slow reaction between buffer layer and the solder. Since the buffer layer is very thin, the Sn solder can diffuse into Cu in a short time. The research work by Yu *et al.* (Yu et al., 2009) reports a 50nm Ni can be used as buffer layer during Cu/Sn/In eutectic bonding. A thin layer of Au has also been used for wetting and for metal layer surface protection from oxidation. During bonding, Sn and In will first wet and react with Ni layer. Alloy of Ni_3Sn_4 or NiInSn ternary phase is formed initially. Then, InSn solder starts to diffuse into Cu to form $Cu_6(Sn,In)_5$ compounds. Finally, all Ni atoms diffuse into $Cu_6(Sn,In)_5$ to form $(Cu,Ni)_6(Sn,In)_5$ phase, and $Au(In,Sn)_2$ is formed as a byproduct (Fig. 16). In addition, the TCT test is a very important examination for the reliability of eutectic bonding technology, since the compound has the potential risk of structural degradation caused by solder fatigue after a long term dramatic change in temperature or brittleness of the inter-metallic compound at low temperature.

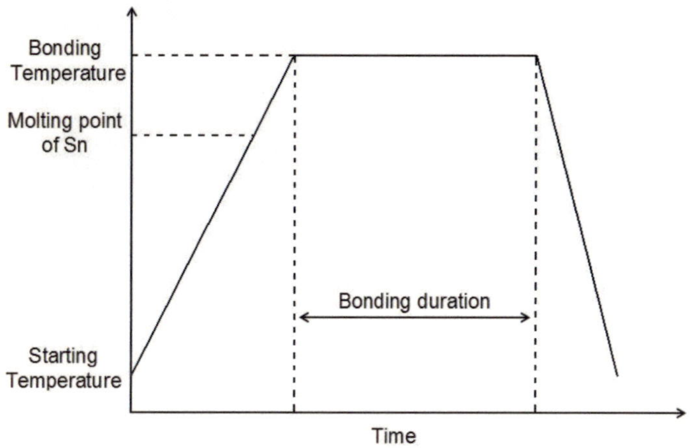

Figure 14. Schematic illustration of the temperature profile during eutectic bonding.

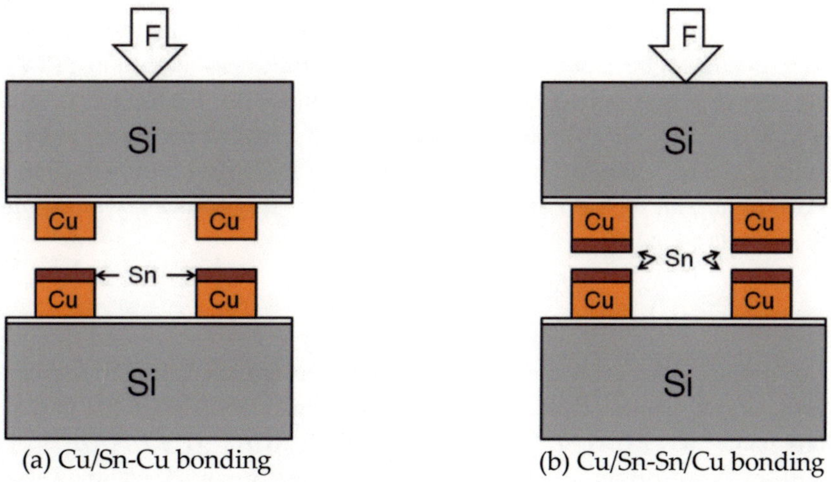

Figure 15. Different methods used in Cu/Sn eutectic bonding.

Good die shear strength and outstanding hermeticity have been obtained using Ni as the buffer layer. The average bonding strength can go up to 32 MPa and the helium leak rate with the seal ring size of 300 μm is smaller than 5×10^{-8} atm.cm^3/sec which is defined as the reject limit for standard MIL-STD-883E method 1014.9. After temperature cycling test (from -40 °C to 125 °C up to 1,000 cycles) and high humidity storage (85 °C, 85% RH for 1000 hr), the bonding strength still remains above 15 MPa, and over 80% dies can still provide high hermeticity level. The research work by Liu *et al.* (Liu et al., 2011) has reported the resistance of bonded interconnects obtained by Cu/Sn bonding. The bonded interconnect shows the resistance of the order of 100 mΩ, and the excellent bonding strength of about 45 Mpa.

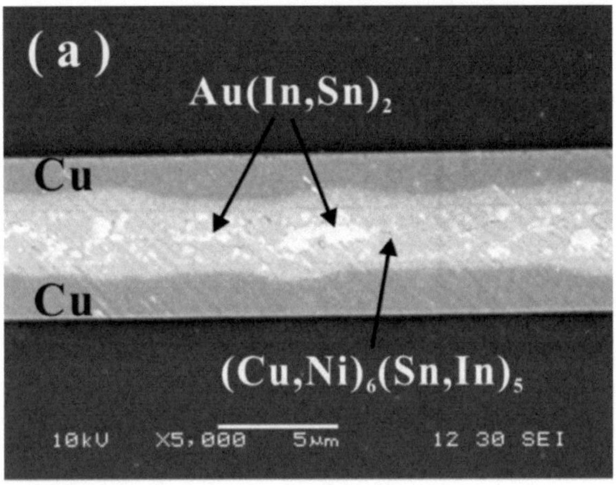

Figure 16. Interfacial microstructure of intermetallic compound joint (Yu et al., 2009) [Copyright of Elsevier].

3.4. Low temperature oxide fusion bonding

Oxide fusion bonding describes the direct bonding between wafers with or without dielectric layers. This bonding method has a stringent surface quality requirement, e.g. wafer surface needs to be smooth with small total thickness variation (TTV) and low roughness is also strictly required. Surface activation is usually performed before bonding. Subsequently, a spontaneous adhesion is firstly applied between the two wafers. Post-bond annealing allows the bonding interface to convert from hydrogen bonds to strong covalent bonds. Surface treatment method and post-bonding annealing process will be presented in detail in the following sections.

3.4.1. Surface activation before bonding

A number of surface activation methods have been investigated, include oxygen plasma bombardment, argon sputter-cleaning, and wet chemical methods with various reagent combinations such as RCA1 (H_2O+H2O_2+NH_4OH), RCA2 (H_2O+H_2O_2+HCl), piranha (H_2SO_4+H_2O_2), etc. The original bonding surface is usually covered with a thin layer of native oxide and contaminant. When the surface is exposed to plasma or immersed in chemical solution, the bombardment of energetic particles or the corrosion of ions removes the surface contamination. At the same time, a very thin high hydrophilic amorphous oxide layer can be formed. Following that, clean and activated surfaces are ready for subsequent hydrophilic bonding. Finally, hydrogen bonds are formed when the two surfaces are brought into contact.

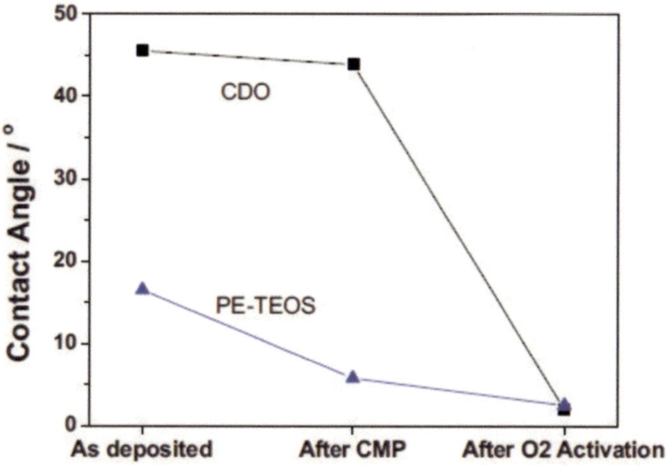

Figure 17. CA measurement for dielectric materials of PE-TEOS and low-*k* CDO under three different conditions, i.e., as deposited, polished, and O_2 plasma activated (Tan&Chong, 2010).

Fig. 17 shows the water droplet contact angle (CA) value of both plasma-enhanced tetraethyl orthosilicate (PE-TEOS) and carbon-doped oxide (CDO) samples for three different conditions, i.e., as deposited, after CMP, and after O_2 plasma activation with CMP. For hydrophilic wafer bonding, smaller contact angle corresponds to higher hydrophilicity of a surface, hence higher density of hydroxyl (OH) groups for hydrogen bond formation during bonding. With O_2 plasma surface activation, both PE-TEOS and CDO show a convergence of CA values to ~2.5°, resulting in a highly hydrophilic surface for fusion bonding.

3.4.2. Post-bonding anneal

For hydrophilic bonding, when two cleaned and activated wafers are brought into contact at room temperature, hydrogen bonds between hydroxyl (–OH) groups are established across the gap between the wafers. Anneal process must be applied after bonding in order to achieve a much higher bonding strength by converting the hydrogen bonds into a strong covalent bonds. The reaction of surface silanol (Si-OH) groups is enhanced during annealing based on the following equation and therefore more covalent bonds are formed:

$$Si - OH + OH - Si \rightarrow Si - O - Si + H_2O \quad (5)$$

For some high-*k* dielectric materials bonding, such as Al_2O_3, HfO_2, and TiO_2 which are used to achieve a higher bonding strength for a given anneal temperature and duration, their reaction during anneal can be presented as follows:

$$M - OH + HO - M \rightarrow M - O - M + H2O \tag{6}$$

Where, M is the symbol for metal atom in high-*k* materials.

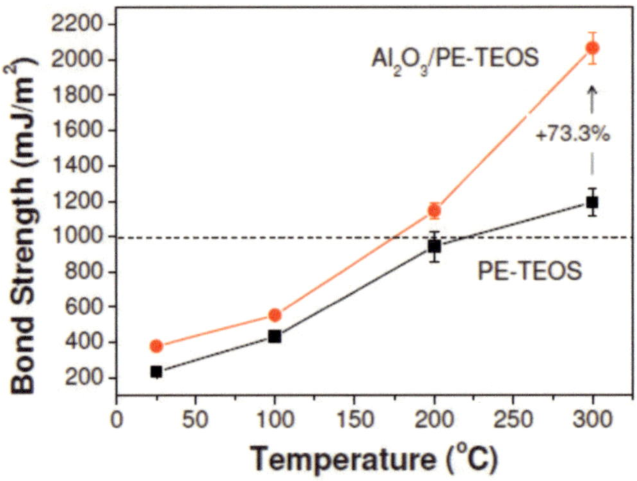

Figure 18. Variation in bond strength of PE-TEOS and Al_2O_3/PE-TEOS samples at various annealing temperatures (Chong&Tan, 2011).

Fig. 18 shows the variation in the bond strength of bonded wafers at various bonding temperatures as measured using the Maszara's crack opening method. The annealing duration is 3 hr. As expected, a higher bonding strength is achieved at a higher annealing temperature. Bonding strength is marginally improved for anneal temperature below 100°C. As the anneal temperature is increased to 200°C, significant improvements in the bonding strength are obtained. When the bonding temperature reaches 300°C, all samples present a bonding strength superior to 1 J/m2 which is the minimum strength required to sustain post-bonding processes such as mechanical grinding and tetramethylammonium hydroxide (TMAH) etching (Tan&Reif, 2005). The enhancement in the bond strength value using a thin Al_2O_3 layer is most likely related to the different bond dissociation energy between Al-O-Al and Si-O-Si. Since the Si-O bond has lower bond dissociation energy (316 kJ/mol) compared with that of the Al-O bond (511 kJ/mol) at 298 K, a higher energy is required to debond wafers that are bonded with Al_2O_3.

4. Summary and conclusion

Over the past decades bonding technology has been used as the mainstream 3D integration method by various key players in America, Asia and Europe. Metal diffusion bonding and eutectic bonding are widely chosen for stacking of multiple chip layers in 3D integration as these methods allow simultaneous formation of mechanical, electrical and hermetic bonds. Although wafer-level stacking using via-first and face-to-face or face-to-back stacking method by Cu diffusion bonding technology has already been investigated with some success, high temperature (>300 °C) processing still remains a challenge. Meanwhile, a number of research work on Cu/Sn eutectic bonding using chip-level stacking method have been demonstrated and provided high vertical interconnect density in 3D stacking. However, formation of inter-metallic compound weakens the quality and reliability of the bonds. As the metal based bonding technology can provide electrical contact, mechanical support and hermetic seal in one simultaneous step, low temperature wafer-level Cu diffusion bonding and Cu/Sn eutectic bonding technologies with inter-layer connection technology, such as TSV, present a very attractive prospect for 3D integration.

Technology development in the areas of 3D integration has resulted in a number of attractive stacking methods. In this chapter, fundamental of low temperature metal diffusion bonding and eutectic bonding technology are introduced. Cu diffusion bonding and Cu/Sn eutectic bonding are presented in details. Another bonding technology using oxide fusion between wafer pair is also exhibited, as it is widely use in the semiconductor industry. Excellent performances of these bonding technologies are shown in the chapter. Some details of processes and methodology used in the research work are included as well.

Author details

Ji Fan and Chuan Seng Tan
Nanyang Technological University, Singapore

Acknowledgement

The authors wish to thank the contributions and support for the work presented in this chapter from group members PENG Lan, LIM Dau Fatt, CHONG Gang Yih, and MADE Riko I in NTU.

5. References

Beyne, E. (2006). 3D system integration technologies. *Proceeding of 2006 International Symposium on VLSI Technology, Systems, and Applications*, Hsinchu, IEEE, pp.19-27.

Chen, K. N., Fan, A., Tan, C. S.&Reif, R. (2004). Contact Resistance Measurement of Bonded Copper Interconnects for Three-dimensional Integration Technology, *IEEE Electr. Device L.* 25(1): 10-12.

Chong, G. Y.&Tan, C. S. (2009). Low Temperature PE-TEOS Oxide Bonding Assisted by a Thin Layer of High-kappa Dielectric, *Electrochem. Solid-State Lett.* 12(11): H408-H411.

Chong, G. Y.&Tan, C. S. (2011). PE-TEOS Wafer Bonding Enhancement at Low Temperature with a High-kappa Dielectric Capping Layer of Al(2)O(3), *J. Electrochem. Soc.* 158(2): H137-H141.

Fan, J., Lim, D. F., Peng, L., Li, K. H.&Tan, C. S. (2011). Low Temperature Cu-to-Cu Bonding for Wafer-Level Hermetic Encapsulation of 3D Microsystems, *Electrochem. Solid-State Lett.* 14(11): H470-H474.

Gibbons, J. F.&Lee, K. F. (1980). "One-Gate-Wide CMOS Inverter on Laser-Recrystallized Polysilicon, *Electron Device Letters* 1(6): 117-118.

Goeloe, G. T., Maby, E. W., Silversmith, D. J., Mountain, R. W.&Antoniadis, D. A. (1981). Vertical Single Gate CMOS Inverters on Laser-Processed Multilayer Substrates.*Proceeding of IEEE International Electron Devices Meeting*, Washington, DC, USA, IEEE, pp.554-556.

Huang, Z. Y., Suo, Z., Xu, G. H., He, J., Prevost, J. H.&Sukumar, N. (2005). Initiation and Arrest of an Interfacial Crack in a Four-point Bend Test, *Eng. Fract. Mech.* 72(17): 2584-2601.

Jang, E. J., Hyun, S., Lee, H. J.&Park, Y. B. (2009). Effect of Wet Pretreatment on Interfacial Adhesion Energy of Cu-Cu Thermocompression Bond for 3D IC Packages, *J. Electro. Mater.* 38(12): 2449-2454.

Jourdain, A., Soussan, P., Swinnen, B.&Beyne, E. (2009). Electrically Yielding Collective Hybrid Bonding for 3D Stacking of ICs.*Proceeding of 59th Electronic Components and Technology Conference*, San Diego, CA, USA, IEEE, pp.11-13.

Jourdain, A., Stoukatch, S., De Moor, P., Ruythooren, W., Pargfrieder, S., Swinnen, B., Beyne, E.&Ieee (2007). Simultaneous Cu-Cu and Compliant Dielectric Bonding for 3D Stacking of ICs.*Proceeding of IEEE International Interconnect Technology Conference*, Burlingame, CA IEEE, pp.207-209.

Kanellos, M. (2003). Moore's Law to Roll on for Another Decade. CNET News Available: http://news.cnet.com/2100-1001-984051.html. Accessed 2003 February 10.

Kim, B., Matthias, T., Cakmak, E., Jang, E. J., Kim, J. W.&Park, Y. B. (2010). Interfacial Properties of Cu-Cu Direct bonds for TSV Integration, *Solid State Technol.* 53(8): 18-21.

Ko, C. T.&Chen, K. N. (2010). Wafer-level Bonding/stacking Technology for 3D Integration, *Microelectron. Reliab.* 50(4): 481-488.

Lapadatu, A., Simonsen, T. I., Kittilsland, G., Stark, B., Hoivik, N., Dalsrud, V.&Salomonsen, G. (2010). Cu-Sn Wafer Level Bonding for Vacuum Encapsulation of Microbolometers Focal Plane Arrays.*Proceeding of Semiconductor Wafer Bonding 11: Science, Technology, and Applications - In Honor of Ulrich Gosele - 218th ECS Meeting*, Las Vegas, NV, United states, pp.73-82.

Lau, J., Lee, C., Premachandran, C.&Yu, A. (2009). *Advanced MEMS Packaging*. New York McGraw-Hill.

Liu, H., Salomonsen, G., Wang, K. Y., Aasmundtveit, K. E.&Hoivik, N. (2011). Wafer-Level Cu/Sn to Cu/Sn SLID-Bonded Interconnects With Increased Strength, *IEEE T. Compon. Pack. T.* 1(9): 1350-1358.

McMahon, J. J., Lu, J. Q., Gutmann, R. J.&Ieee (2005). Wafer Bonding of Damascene-patterned Metal/Adhesive Redistribution Layers for Via-first Three-dimensional (3D) Interconnect.*Proceeding of 55th Electronic Components & Technology Conference*, Lake Buena Vista, FL, USA, IEEE, pp.331-336.

Moore, G. E. (1998). Cramming more components onto integrated circuits (Reprinted from Electronics, pg 114-117, April 19, 1965), *P. IEEE.* 86(1): 82-85.

Munding, A., Hubner, H., Kaiser, A., Penka, S., Benkart, P.&Kohn, E. (2008). Cu/Sn Solid-liquid Interdiffusion Bonding, *Wafer Level 3-D ICs Process Technology*, Springer.

Nadipalli, R., Fan, J., Li, K. H., Wee, K. W., Yu, H.&Tan, C. S. (2012). 3D Integration of MEMS and CMOS via Cu-Cu Bonding with Simultaneous Formation of Electrical, Mechanical and Hermetic Bonds.*Proceeding of IEEE International 3D System Integration Conference (3DIC)*, Osaka, Japan, IEEE

Niklaus, F., Enoksson, P., Kalvesten, E.&Stemme, G. (2001). Low-temperature full wafer adhesive bonding, *J. Micromech. Microeng.* 11(2): 100-107.

Pan, C. T., Yang, H., Shen, S. C., Chou, M. C.&Chou, H. P. (2002). A Low-Temperature Wafer Bonding Technique Using Patternable Materials, *J. Micromech. Microeng.* 12(5): 611-615.

Pavlidis, V. F.&Friedman, E. G. (2009). *Three-dimensional Integrated Circuit Design*, Elsevier.

Peng, L., Li, H. Y., Lim, D. F., Gao, S.&Tan, C. S. (2011). High-Density 3-D Interconnect of Cu-Cu Contacts With Enhanced Contact Resistance by Self-Assembled Monolayer (SAM) Passivation, *IEEE Trans. Electron Devices* 58(8): 2500-2506.

Tadepalli, R.&Thompson, C. V. (2003). Quantitative Characterization and Process Optimization of Low-Temperature Bonded Copper Interconnects for 3-D Integrated Circuits.*Proceeding of IEEE International Interconnect Technology Conference*, Burlingame, CA, USA, IEEE, pp.30-38.

Tan, C. S.&Chong, G. Y. (2010). Low Temperature Wafer Bonding of Low-kappa Carbon-Doped Oxide for Application in 3D Integration, *Electrochem. Solid-State Lett.* 13(2): H27-H29.

Tan, C. S.&Reif, R. (2005). Microelectronics Thin Film Handling and Transfer Using Low-temperature Wafer Bonding, *Electrochem. Solid-State Lett.* 8(12): G362-G366.

Tan, C. S., Fan, J., Lim, D. F., Chong, G. Y.&Li, K. H. (2011). Low Temperature Wafer-level Bonding for Hermetic Packaging of 3D Microsystems, *J. Micromech. Microeng.* 21(7): 075006.

Tan, C. S., Lim, D. F., Ang, X. F., Wei, J.&Leong, K. C. (2012). Low Temperature Cu-Cu Thermo-compression Bonding with Temporary Passivation of Self-assembled Monolayer and Its Bond Strength Enhancement, *Microelectron. Reliab.* 52(2): 321-324.

Tan, C. S., Lim, D. F., Singh, S. G., Goulet, S. K.&Bergkvist, M. (2009). Cu-Cu Diffusion Bonding Enhancement at Low Temperature by Surface Passivation Using Self-assembled Monolayer of Alkane-thiol, *Appl. Phys. Lett.* 95(19).

Whelan, C. M., Kinsella, M., Carbonell, L., Ho, H. M.&Maex, K. (2003). Corrosion Inhibition by Self-assembled Monolayers for Enhanced Wire Bonding on Cu Surfaces, *Microelectron. Eng.* 70(2-4): 551-557.

Xu, D. H., Jing, E. R., Xiong, B.&Wang, Y. L. (2010). Wafer-Level Vacuum Packaging of Micromachined Thermoelectric IR Sensors, *IEEE T. Adv. Pack.* 33(4): 904-911.

Young, A. M.&Koester, S. J. (2010). 3D Process Technology Considerations, *Three-Dimensional Integrated Circuit Design*, Springer: 15-32.

Yu, D. Q., Lee, C., Yan, L. L., Thew, M. L.&Lau, J. H. (2009). Characterization and Reliability Study of Low Temperature Hermetic Wafer Level Bonding Using In/Sn Interlayer and Cu/Ni/Au Metallization, *J. Alloy. Compd.* 485(1-2): 444-450.

Yun, C. H., Brosnihan, T. J., Webster, W. A.&Villarreal, J. (2005). Wafer-level packaging of MEMS accelerometers with through-wafer interconnects.*Proceeding of 55th Electronic Components & Technology Conference*, Lake Buena Vista, FL, USA, IEEE, pp.320-323.

Yun, C. H., Martin, J. R., Tarvin, E. B.&Winbigler, J. T. (2008). Al to Al Wafer Bonding for MEMS Encapsulation and 3-D Interconnect.*Proceeding of IEEE International Conference on Microelectromechanical Systems (MEMS 2008)*, Tucson, AZ, USA, IEEE, pp.810-813.

Chapter 5

Homogenization Heat Treatment to Reduce the Failure of Heat Resistant Steel Castings

Mohammad Hosein Bina

Additional information is available at the end of the chapter

http://dx.doi.org/10.5772/50312

1. Introduction

In this chapter, influence of the homogenization heat treatment on the failure of heat resistant steel castings is studied. First, the cast stainless steels and effective factors on the failure of these steels are described. One of the most important factors is sigma phase embrittlement which is studied, in detail. Finally, the effect of homogenization heat treatment on dissolution of carbides and reduction of sigma-phase and failure is discussed. The present chapter is a compilation of my experience in industry and other studies about fracture of continuous annealing furnace rollers, prepared for use by practitioners and researchers. The chapter may also be useful for graduate students, researching failure.

2. Classification and designation of cast stainless steels

Cast stainless steels are usually classified as either corrosion resistant steel castings (which are used in aqueous environments below 650°C) or heat resistant steel castings (which are suitable for service temperatures above 650 °C). Cast stainless steels are most often specified on the basis of chemical composition using the designation system of the High Alloy Product Group of the Steel Founders Society of America [1].

2.1. Corrosion resistant steel castings

The corrosion resistant steel castings are widely used in chemical processes and electricity equipment that need corrosion resistant in aqueous or liquid-vapor environments at the temperatures less than 315°C. The serviceability of cast corrosion-resistant steels depends greatly on the absence of carbon, and especially precipitated carbides, in the alloy microstructure. Therefore, cast corrosion-resistant alloys are generally low in carbon (usually lower than 0.20% and sometimes lower than 0.03%). All cast corrosion-resistant steels contain more than 11% chromium [1].

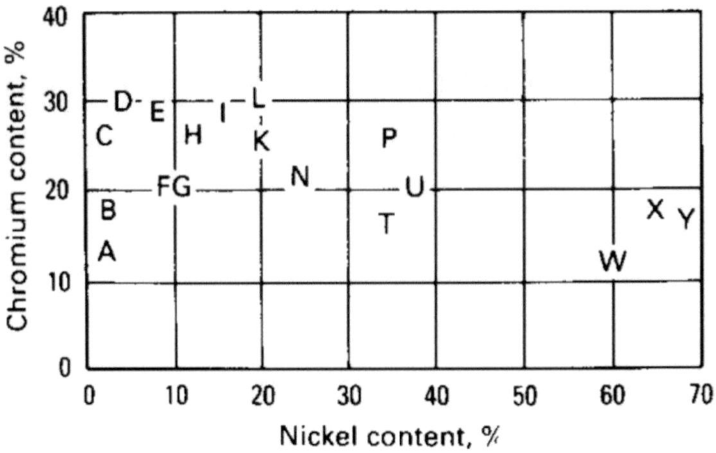

Figure 1. Chromium and nickel contents in ACI standard grades of heat- and corrosion-resistant steel castings [1].

2.2. Heat resistant steel castings

Stainless steel castings are classified as heat resistant if they are capable of sustained operation while exposed, either continuously or intermittently, to operating temperatures that result in metal temperatures in excess of 650°C. Heat resistant steel castings resemble high-alloy corrosion resistant steels except for their higher carbon content, which imparts greater strength at elevated temperature [1]. The three principal categories of H-type cast steels, based on composition, are [1–4]:

a. Iron-chromium alloys containing 10 to 30% Cr and little or no nickel. These alloys have low strength at elevated temperatures and are useful mainly due to their resistance to oxidation. Use of these alloys is restricted to conditions, either oxidizing or reducing, that involve low static loads and uniform heating. Chromium content depends on anticipated service temperature [1].
b. Iron-chromium-nickel alloys contain more than 13% Cr and 7% Ni (always more chromium than nickel) [1]. These austenitic alloys are ordinarily used under oxidizing or reducing conditions similar to those withstood by the ferritic iron-chromium alloys, but in service they have greater strength and ductility than the straight chromium alloys. They are used, therefore, to withstand greater loads and moderate changes in temperature. These alloys also are used in the presence of oxidizing and reducing gases that are high in sulfur content.
c. Iron-nickel-chromium alloys contain more than 25% Ni and more than 10% Cr (always more nickel than chromium) [1]. These austenitic alloys are used for withstanding reduction as well as oxidizing atmospheres, except where sulfur content is appreciable. (In atmospheres containing 0.05% or more hydrogen sulfide, for example, iron-chromium-nickel alloys are recommended [1].) In contrast with iron-chromium-nickel alloys, iron-nickel-chromium alloys do not carburize rapidly or become brittle and do not take up nitrogen in nitriding atmospheres. These characteristics become enhanced

as nickel content is increased, and in carburizing and nitriding atmospheres casting life increases with nickel content [1]. Austenitic iron-nickel-chromium alloys are used extensively under conditions of severe temperature fluctuations such as those encountered by fixtures used in quenching and by parts that are not heated uniformly or that are heated and cooled intermittently. In addition, these alloys have characteristics that make them suitable for electrical resistance heating elements.

3. Microstructure of heat resistant steels

The microstructure of a particular grade is primarily determined by composition. Chromium, molybdenum, and silicon promote the formation of ferrite (magnetic), while carbon, nickel, nitrogen, and manganese favor the formation of austenite (nonmagnetic). Chromium (a ferrite and martensite promoter), nickel, and carbon (austenite promoters) are particularly important in determining microstructure [1]. The heat resistant casting alloys (except HA-type) are contain high proportions of chromium and nickel that are generally austenitic and nonmagnetic. The austenite in the matrix provides useful high-temperature strength if it is adequately reinforced with particles of carbide and nitride. The austenite must contain no ferrite to reach maximum strength [5]. The microstructure of HH-type austenitic stainless steel castings (25Cr-12Ni) used in continuous annealing furnace rollers in un-worked condition is presented in Fig. 2. The ideal microstructure of HH alloy is

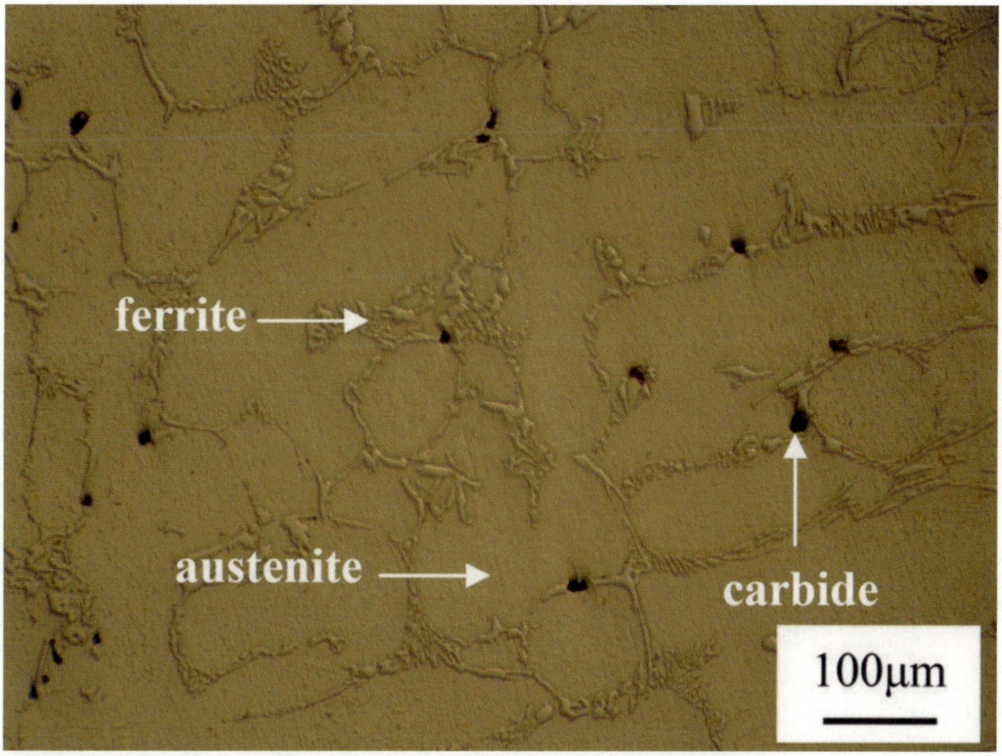

Figure 2. Microstructure of roller at un-worked condition (2% nital agent).

austenitic, while, in Fig. 2, the initial microstructure consisted of austenite, ferrite, and also black carbide particles. The presence of ferrite was attributed to un-controlled solidification and therefore, occurrence segregation phenomenon in the sample.

The presence of ferrite in microstructure may be beneficial or detrimental (especially at high temperatures), depending on the application. Ferrite can be beneficial in terms of weldability because fully austenitic stainless steels are susceptible to a weldability problem known as hot cracking, or microfissuring. But, at 540°C and above, the ferrite phase may transform to a complex iron-chromium-nickel-molybdenum intermetallic compound known as σ phase, which reduces toughness, corrosion resistance, and creep ductility [1]. The extent of the reduction in strength increases with time and temperature to about 815°C and may persist to 925°C [1].

4. Sigma phase embrittlement

Several factors cause failure of heat resistant steel castings such as oxidation, sulfidation, carburization, creep, thermal fatigue, and sigma phase embrittlement. Intermetallics such as the sigma phase are important sources of failure in high-temperature materials.

The existence of σ phase in iron-chromium alloys was first detected in 1907 by the observation of a thermal arrest in cooling curves [6]. The first actual observation of σ in iron-chromium alloys was reported in 1927 [7]. The σ phase was identified by x-ray diffraction in 1927 [8] and in 1931 [9]. After the existence of σ was firmly established, numerous studies were conducted to define the compositions and temperatures over which σ could be formed. In general, σ forms with long-time exposure in the range of 565 to 980°C, although this range varies somewhat with composition and processing [1]. Sigma phase has a tetragonal crystal structure with 30 atoms per unit cell and a *c*/*a* ratio of approximately 0.52 [10]. Sigma also forms in austenitic alloys. In fully austenitic alloys, σ forms from the austenite along grain boundaries. If δ-ferrite is present in the austenitic alloy, σ formation is more rapid and occurs in the δ-ferrite [1].

This phase is a unique combination of iron and chromium that produces a hard and brittle second phase. The presence of sigma phase not only results in harmful influence on the mechanical properties of the material, but also reduces its corrosion resistance by removing chromium and molybdenum from the austenitic matrix. A relatively small amount of sigma phase, when it is nearly continuous at a grain boundary, can lead to the very early failure of parts [11]. Fig. 3(a) shows a failed roller (HH-type) which was used for metal strip transfer in continuous annealing furnaces. An austenitic matrix and a network of sigma phase precipitates on austenitic grain boundaries in the failed roller is shown in Fig. 3(b). Formation of the sigma phase during prolonged heating at temperatures exceeding 650°C, results in a detrimental decrease in roller toughness at lower temperatures and during shut-down periods. Therefore, during shut-down, straightening or heating-up periods, the crack propagation along sigma phase at grain boundaries result in failure of rollers [12,13].

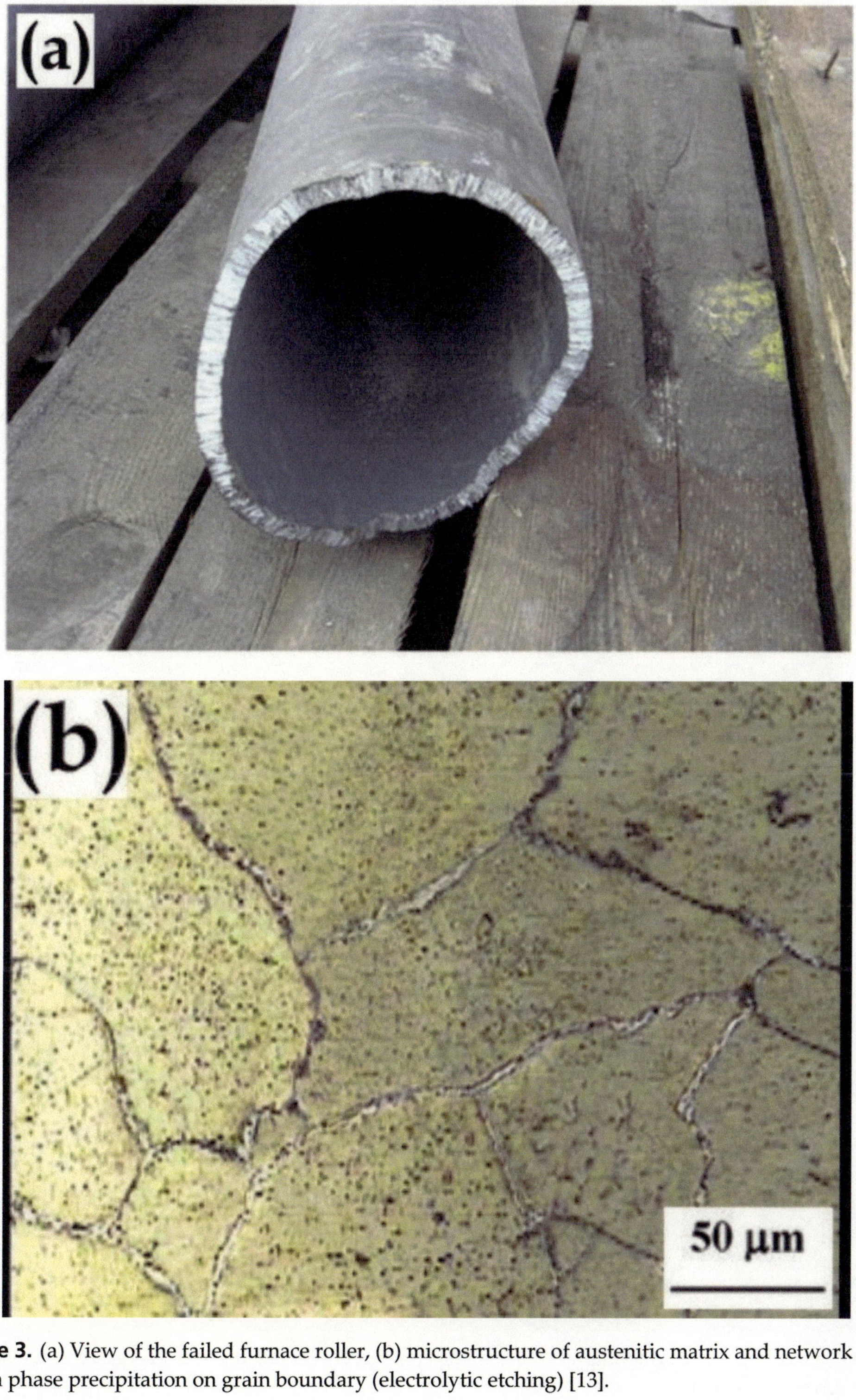

Figure 3. (a) View of the failed furnace roller, (b) microstructure of austenitic matrix and network of sigma phase precipitation on grain boundary (electrolytic etching) [13].

All of the ferritic stabilizing elements promote σ formation [1]. In commercial alloys, silicon, even in small amounts, markedly accelerates the formation of σ. In general, all of the elements that stabilize ferrite promote σ formation. Molybdenum has an effect similar to that of silicon, while aluminum has a lesser influence. Increasing the chromium content, also favors σ formation. Small amounts of nickel and manganese increase the rate of σ formation, although large amounts, which stabilize austenite, retard σ formation. Carbon additions decrease σ formation by forming chromium carbides, thereby reducing the amount of chromium in solid solution [14,15]. Additions of tungsten, vanadium, titanium, and niobium also promote σ formation. As might be expected, σ forms more readily in ferritic than in austenitic stainless steels [14–17].

4.1. Formation mechanism of sigma phase in heat resistant steels

Fig. 4 shows the formation mechanism of sigma phase in the failed roller. The presence of delta-ferrite in the austenitic matrix leads to accelerated sigma-phase formation. Generally, this phase nucleates inside delta-ferrite. Transformation of delta-ferrite can be described by two eutectoid reactions [1]:

$$\delta \rightarrow M_{23}C_6 + \gamma_2 \quad (1)$$

$$\delta \rightarrow \sigma + \gamma_2 \quad (2)$$

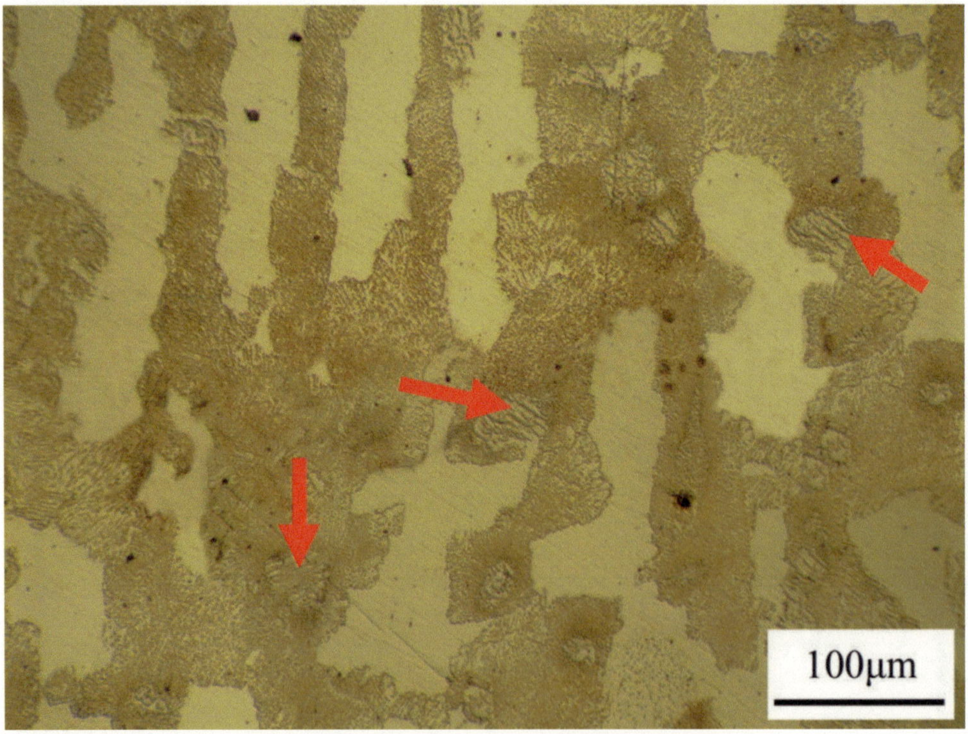

Figure 4. Formation of pearlite-like structure in the cross section of failed roller. The arrows show the large pearlite-like colonies.

By reaction (1), $M_{23}C_6$ carbides with lamellar morphology and also secondary austenite formed (Fig. 4). The mechanism of this reaction can be summarized as follows. first, the growth of carbide precipitates inside delta-ferrite lead to the formation of secondary austenite [18]. The chromium content near precipitates is high and results in increase in the content of chromium close to adjacent delta-ferrite. Therefore, the carbides growth develops inside delta-ferrite and a lamellar structure including carbides and secondary austenite is formed [19]. Finally, after completion of lamellar precipitation, the sigma-phase forms at foreside of the precipitates [19].

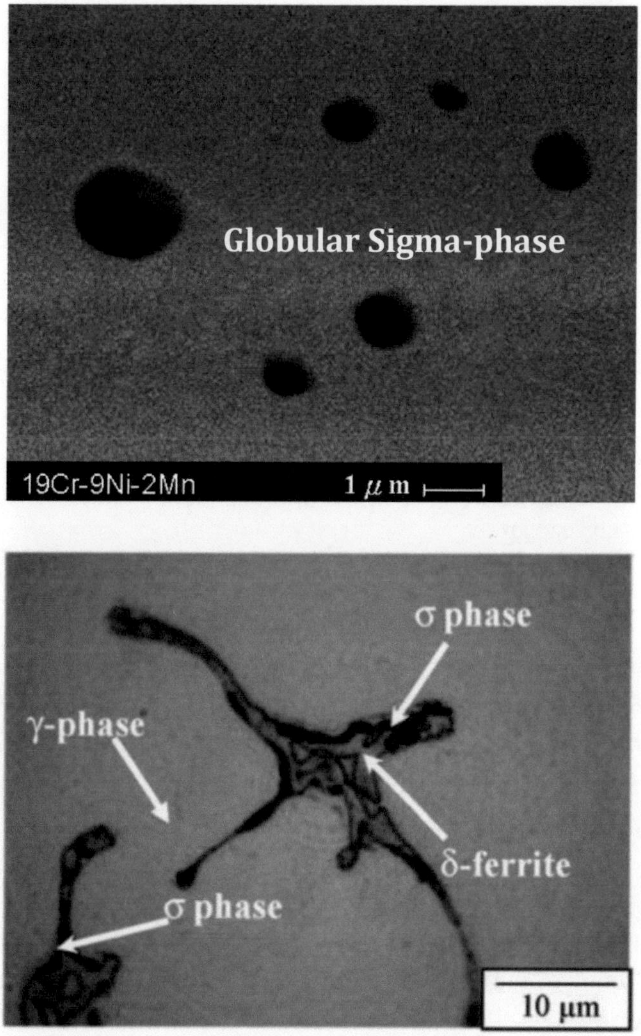

Figure 5. Various morphologies of sigma phase in a stainless steel [20].

4.2. Morphology of sigma phase

The morphologies of sigma-phase can be classified into dendritic and globular structures [17,20,21]. Fig. 5 shows the morphologies of sigma phase in a stainless steel (19Cr-9Ni-2Mn). The sigma-phase was formed by transformation of the delta-ferrite to the embrittling sigma-phase and secondary austenite due to unsuitable working conditions. Lin et al. [20] have pointed out that the dendrite-like sigma-phase is unstable, while the globular sigma-phase is a stable one. Furthermore, dendrite-like σ morphology was observed surrounding the δ-ferrite particles, which meant that the $\delta \rightarrow \sigma$ phase transformation occurred partially [20]. Gill et al. [21] also have proposed that the spheroid σ phase results from the instability of dendritic sigma-phase to any localized decrease in width.

4.3. Precipitation sites of sigma phase

The precipitation sites of sigma-phase consist of δ/γ interface boundary, triple conjunction, grain corner and cellular which are shown in Fig. 6 [22]. The initial precipitation sites are δ/γ interphase boundary because it has higher boundary energy and many defects are concentrated here. Therefore, the precipitation of sigma-phase takes place preferentially at δ/γ boundary, and then precipitates toward interior of delta-ferrite grain [19,22]. The other precipitation sites were concentrated at δ-ferrite because σ-phase preferred to precipitate at a higher Cr content region [20,22]. The precipitation at grain corner meant that sigma-phase was strongly concentrated and formed at the corner of delta-ferrite. The precipitation at triple conjunction meant the σ-phase precipitated surrounding δ-ferrite. The cellular shape precipitation presented the eutectoid decomposition from delta-ferrite into sigma and secondary austenite phases that can be observed clearly at 800°C [22].

4.4. Effect of sigma phase on mechanical properties

One of the most affected mechanical properties of steels by formation of the sigma phase is impact energy. Effect of sigma phase on the impact energy of austenitic steel Fe-25Cr-20Ni is shown in Fig. 8 [23]. By increasing the time of exposure at formation temperature range of sigma phase (760–870°C), toughness value decreases by 85%.

The influence of high-temperature exposure on the toughness of a low-interstitial 29Cr-4Mo ferritic stainless steel has been examined (Fig. 9) by Aggen et al. [24]. In this figure, the C-curve shows the time for embrittlement as a function of aging temperature. For σ formation, embrittlement was most rapid at about 775°C, whereas 475°C embrittlement was slower with a maximum rate at about 480°C. In general, for this alloy, the sigma-phase forms over the range of 595 to 925°C. Embrittlement is most pronounced when intergranular σ films form, resulting in intergranular tensile and impact fractures [1].

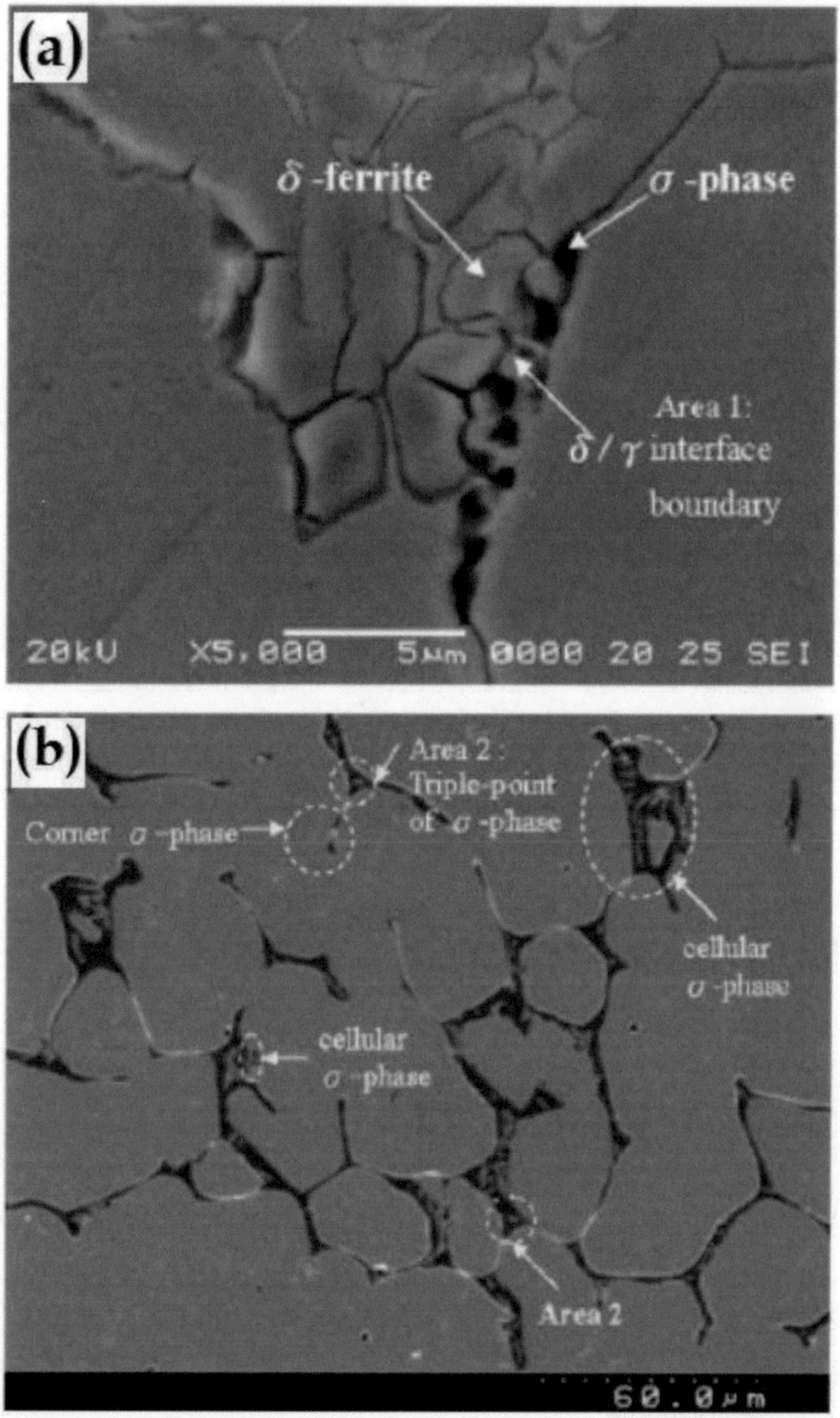

Figure 6. Precipitation of sigma-phase at: (a) δ/γ interface boundary, and (b) triple conjunction, grain corner and cellular [22].

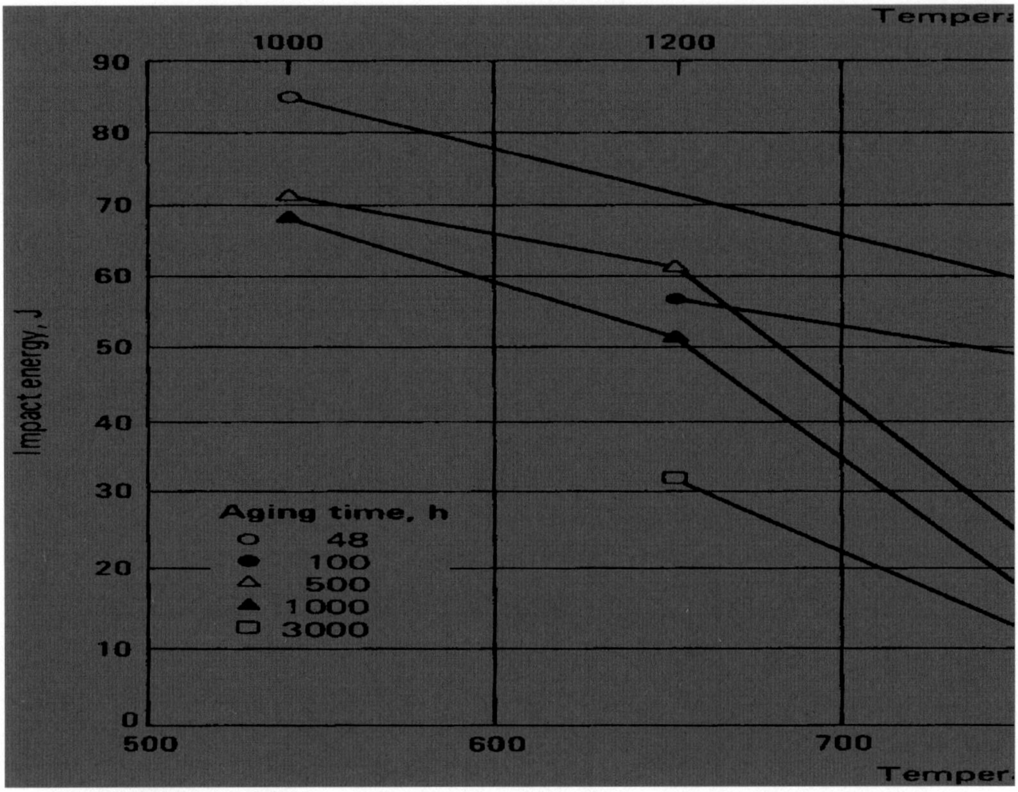

Figure 7. Effect of time and temperature of aging treatment on the impact energy of austenitic steel Fe-25Cr-20Ni [23].

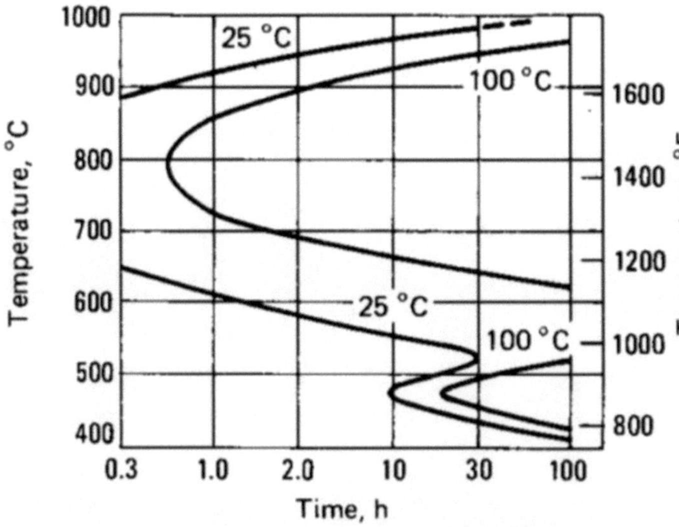

Figure 8. Time-temperature relationships to produce 25 and 100°C DBTTs for a 29Cr-4Mo ferritic stainless steel as a function of aging times that cover both the 475°C embrittlement range and the σ phase embrittlement range [24].

The hardness of sigma phase in Fe-Cr alloys is approximately 68 HRC [1]. Fig. 10 shows the microstructure of austenitic matrix and network of sigma phase precipitation on grain boundary and also indenter effects on the surface at several locations. It can be seen that because of brittleness and higher hardness of sigma phase with respect to the austenite matrix, this phase often fractures during indentation.

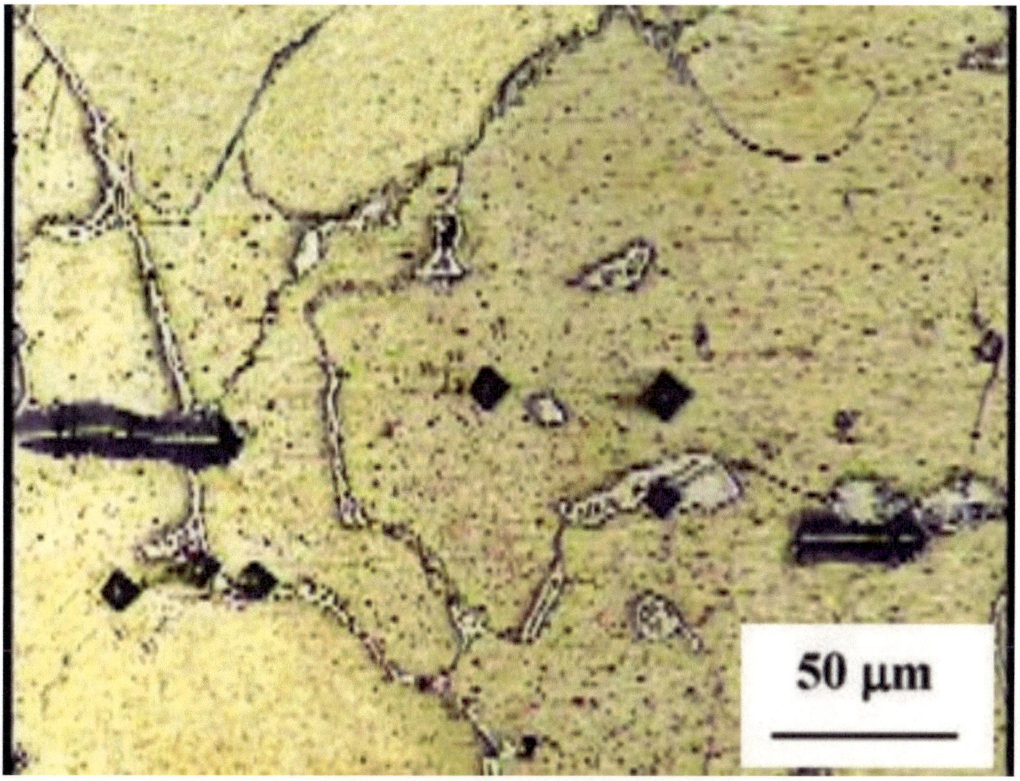

Figure 9. Microstructure of austenitic matrix and network of sigma phase precipitation on grain boundary. It can be seen that because of sigma is brittle phase, often fractures during indentation (indicated by the arrows) [12].

Therefore, a relatively small quantity of the σ phase, when it is nearly continuous at a grain boundary, can lead to very early failure of high-temperature parts [12].

5. Homogenization heat treatment

To eliminate segregation in the cast structure, it is frequently necessary to homogenize the part before usage to promote uniformity of chemical composition and microstructure. This treatment can also be used for complete dissolution of carbides and brittle and deleterious phases like sigma which are formed at operating conditions. In this method, steel is heated for a long enough period of time in order to complete the dissolution of carbides, and then the cooling starts at an appropriate rate to avoid formation of deleterious phases [12].

Specimen	Homogenization temperature (°C)	Cooling environment	Microstructures (etched electrolytically by 10 M KOH solution)	Microstructures (etched by Marble's reagent)
HH1	-	-	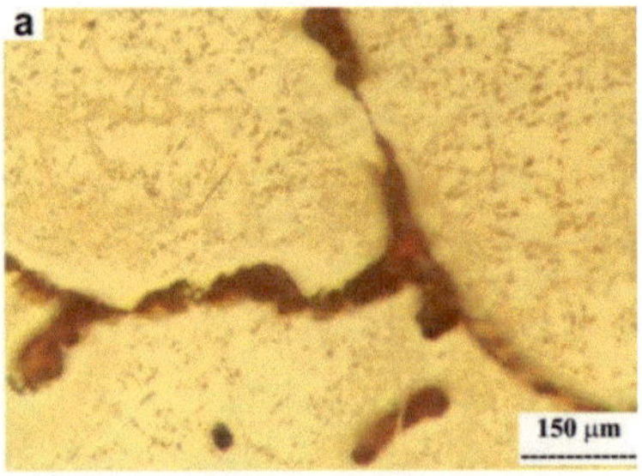	
HH2	950	Air	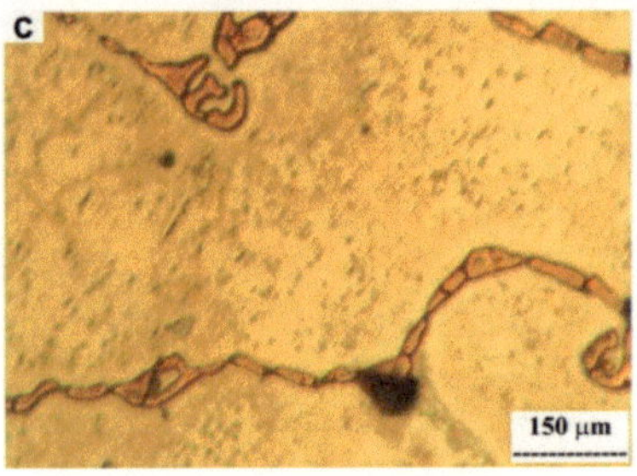	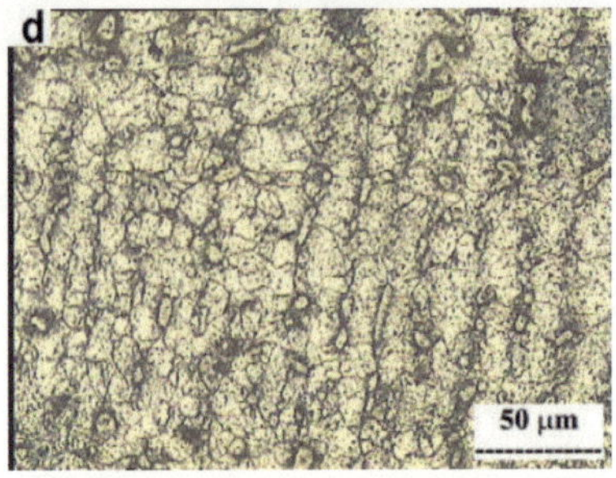
HH8	1050	Air	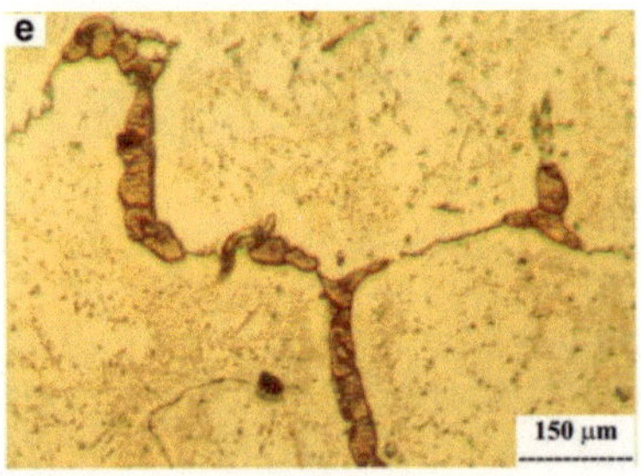	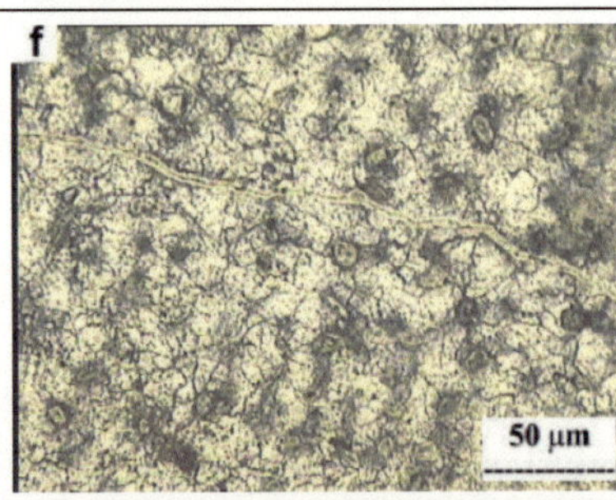
HH11	1100	Air	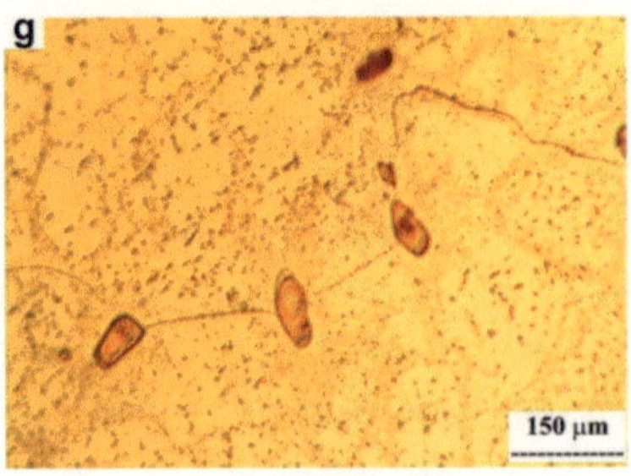	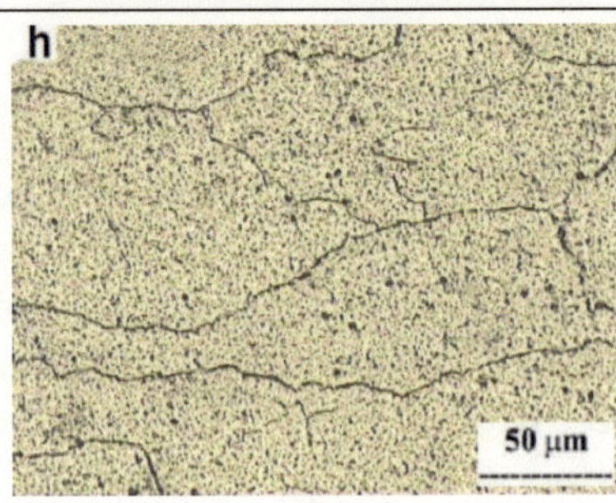

HH14	1100	Furnace	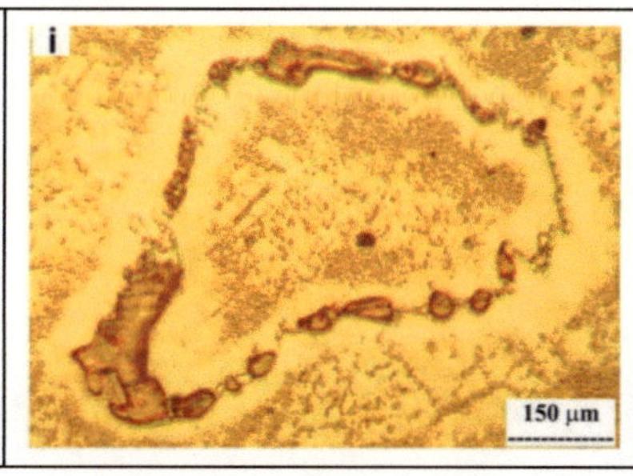	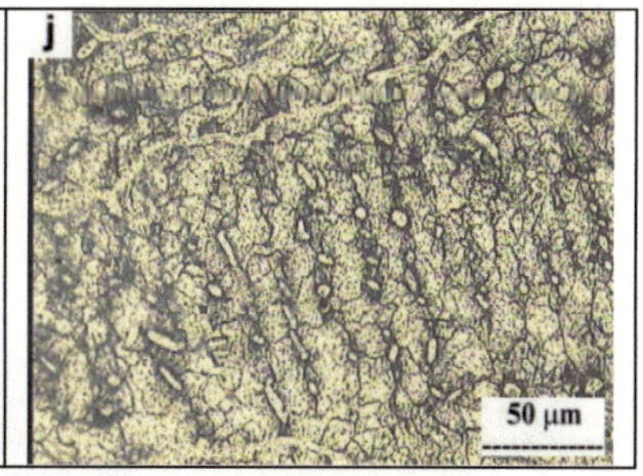

Table 1. Microstructures of selected failed roller steel, before and after the homogenization heat treatment for 2 h.

Microstructures of selected failed roller steel, before and after the homogenization heat treatment at various temperatures for 2 h are shown in Table 1. Formation of the brittle sigma phase and precipitation of carbides is clear in the microstructure of specimen HH14 when compared to the microstructure of specimen HH11. As a result of slow cooling in the furnace, all the desirable changes in the homogenized structure are reversed. In other words, the formation of carbide precipitates in grains, inhomogeneity of austenite as the matrix phase, and the formation of intermetallic brittle phases like sigma phase, which are all results of slow cooling at temperatures between 650 and 950°C, cause the toughness of the steel to decrease greatly. Since the formation of the sigma phase is a time consuming process, formation of this phase has been minimized by cooling in the air. The lowest amounts of sigma phase and carbides are in specimen HH11, which shows homogeneity of microstructure and a decrease in microscopic segregation. Therefore, for roller straightening and a good toughness to be obtained, homogenization heat treatment should be performed at 1100°C for at least 2 h followed by cooling in air [12].

Fig. 11 shows the precipitation sites of sigma-phase in the failed roller before and after the homogenization heat treatment at 1100 °C for 2 h followed by air cooling. Fig. 12 and Table 2 illustrate the chemical composition at three scanning points indicated in Fig. 11 (a), (b) and (c). As can be seen in Fig. 11, the points 1, 2, and 3 were austenite matrix, delta-ferrite, and sigma-phase, respectively. Table 2 shows that the chromium, molybdenum, and silicon content in the sigma-phase were higher than that of the austenite-phase. Generally, delta-ferrite and sigma-phase are Cr-rich. But, the weight percent of silicon and molybdenum in the delta-ferrite is higher than that of other phases (austenite and sigma). Silicon and molybdenum plays an important role in $\delta \rightarrow \sigma + \gamma_2$ phase transformation and acts as strong stabilizers for delta-ferrite [5,20].

As can be seen in Fig. 11, the heat treatment at 1100°C followed by cooling in air led to changing the morphology of sigma phase from dendritic structure (Figs. 11(a) and (b)) to globular structure (Fig. 11(c)). The dendritic sigma phase indicates an unstable shape, and globular sigma phase exhibits a stable shape. Also, the dendritic structure is brittle, while, the globular structure is ductile. Therefore, it can be concluded that for completing the transformation of delta-ferrite to sigma phase, the temperature of homogenization heat treatment should be increased to 1100°C and the morphology of sigma phase be transformed to stable globular structure. This structure leads to remarkable increase in the impact energy and ductility of continuous annealing furnace roller. In contrast, the formation of sigma-phase with dendritic morphology results in decrease in ductility and failure of the roller.

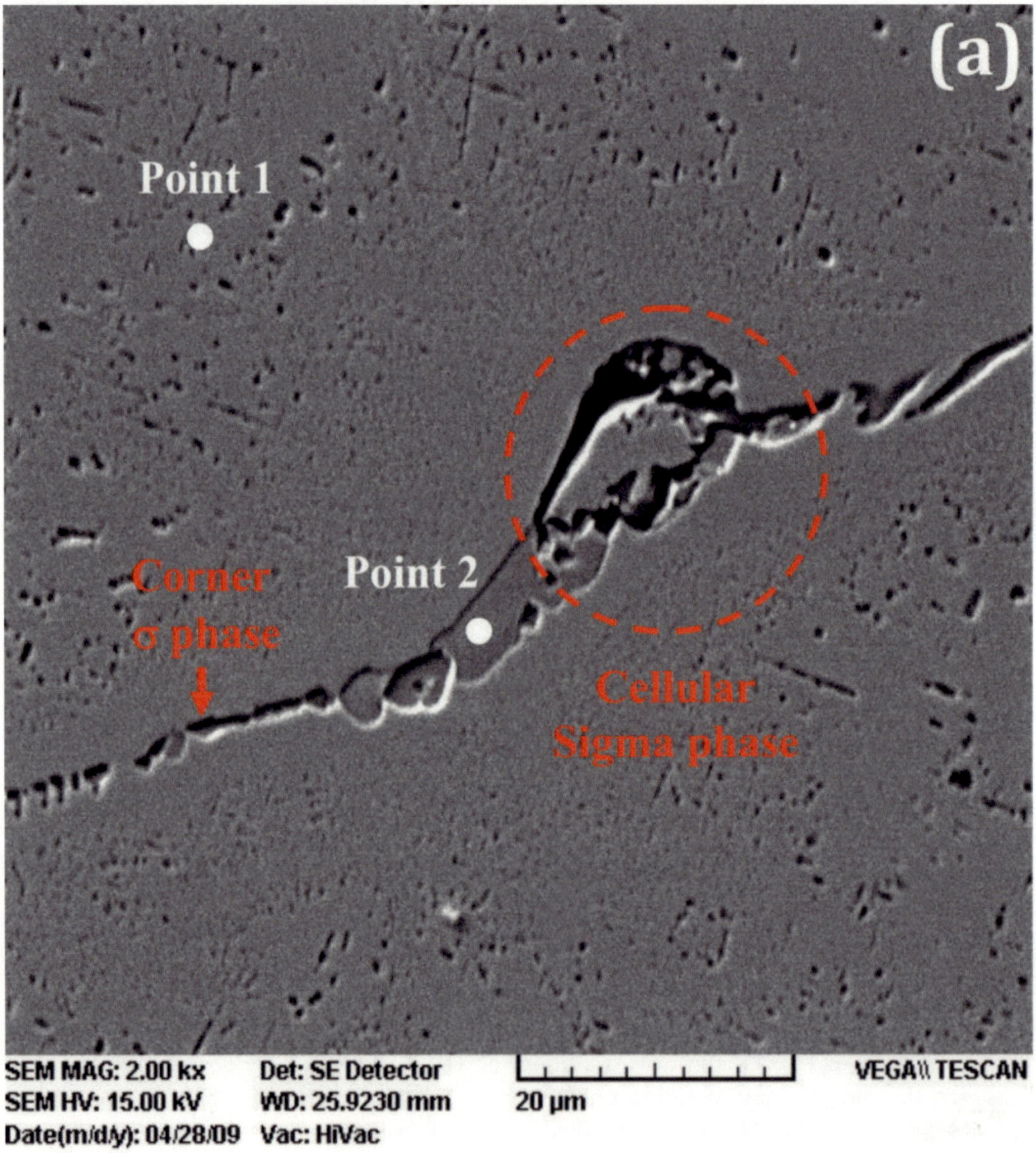

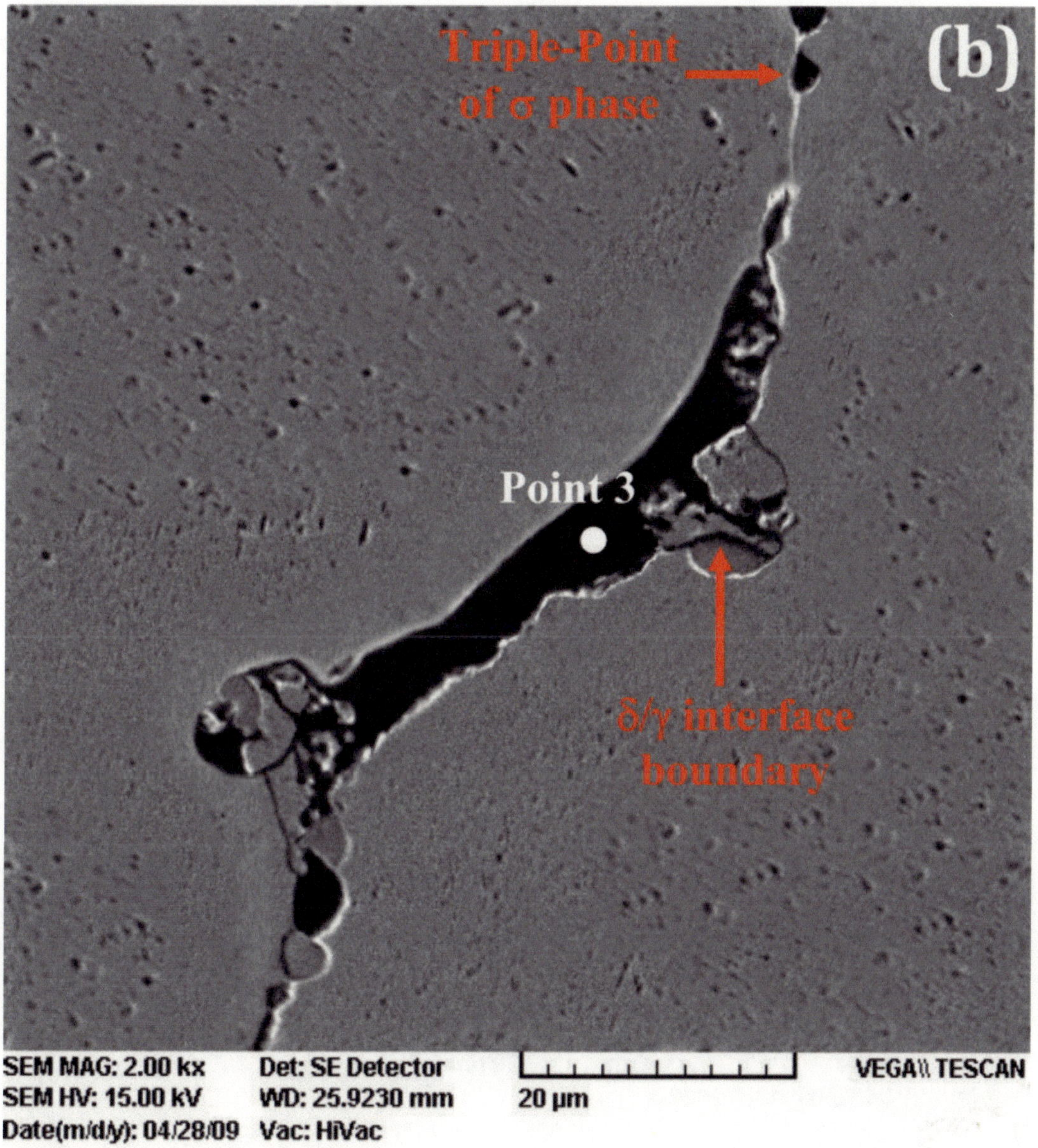
Triple-Point
of σ phase
(b)
Point 3
δ/γ interface
boundary
SEM MAG: 2.00 kx
SEM HV: 15.00 kV
Date(m/d/y): 04/28/09
Det: SE Detector
WD: 25.9230 mm
Vac: HiVac
20 µm
VEGA\\ TESCAN

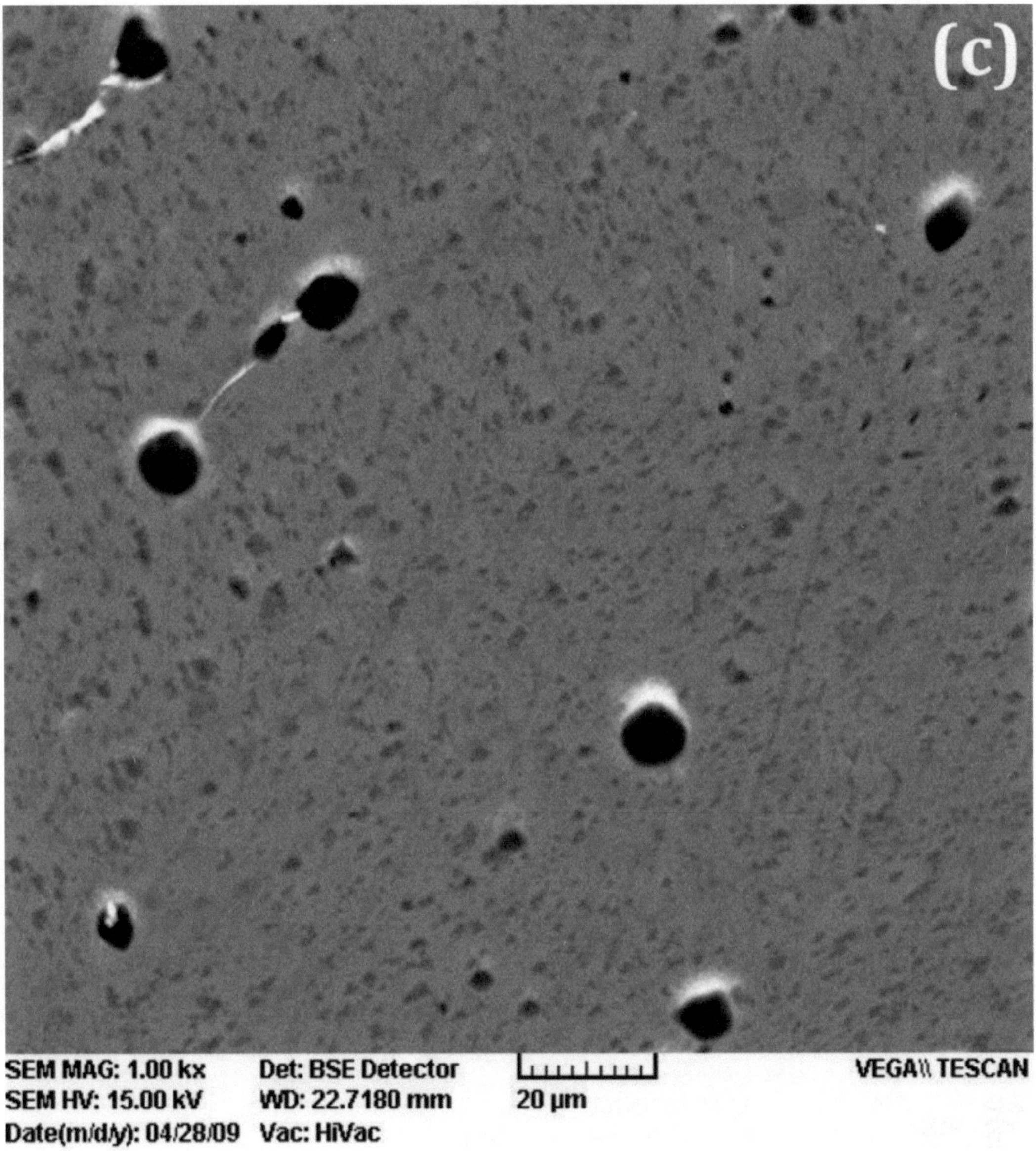

Figure 10. Morphology of sigma-phase. (a) and (b) after failure and before homogenization treatment, respectively and (c) after the homogenization heat treatment at 1100 °C for 2 h followed by air cooling.

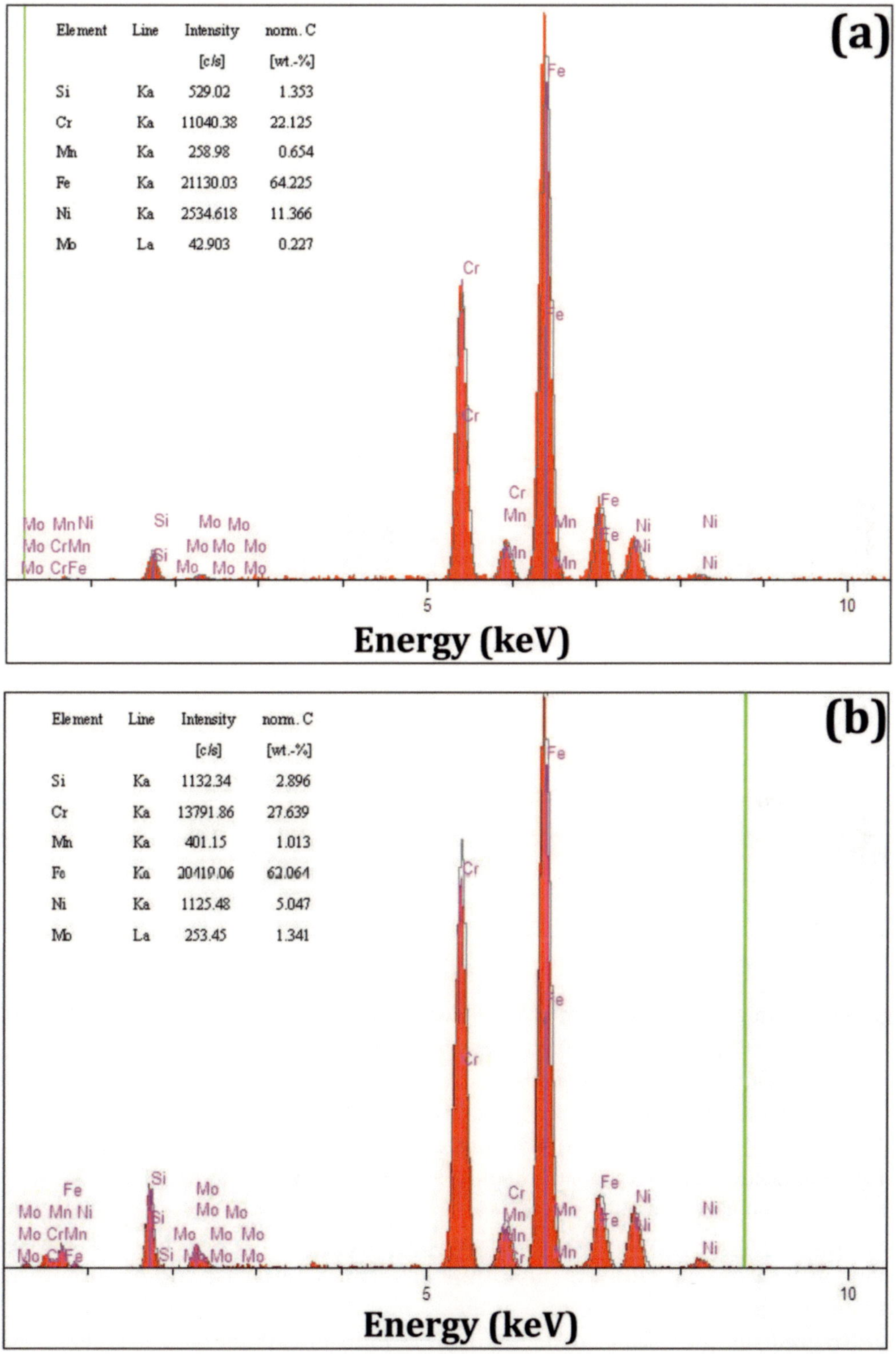
(a)
Element Line Intensity [c/s] norm. C [wt.-%]
Si Ka 529.02 1.353
Cr Ka 11040.38 22.125
Mn Ka 258.98 0.654
Fe Ka 21130.03 64.225
Ni Ka 2534.618 11.366
Mo La 42.903 0.227
5
10
Energy (keV)
(b)
Element Line Intensity [c/s] norm. C [wt.-%]
Si Ka 1132.34 2.896
Cr Ka 13791.86 27.639
Mn Ka 401.15 1.013
Fe Ka 20419.06 62.064
Ni Ka 1125.48 5.047
Mo La 253.45 1.341
5
10
Energy (keV)

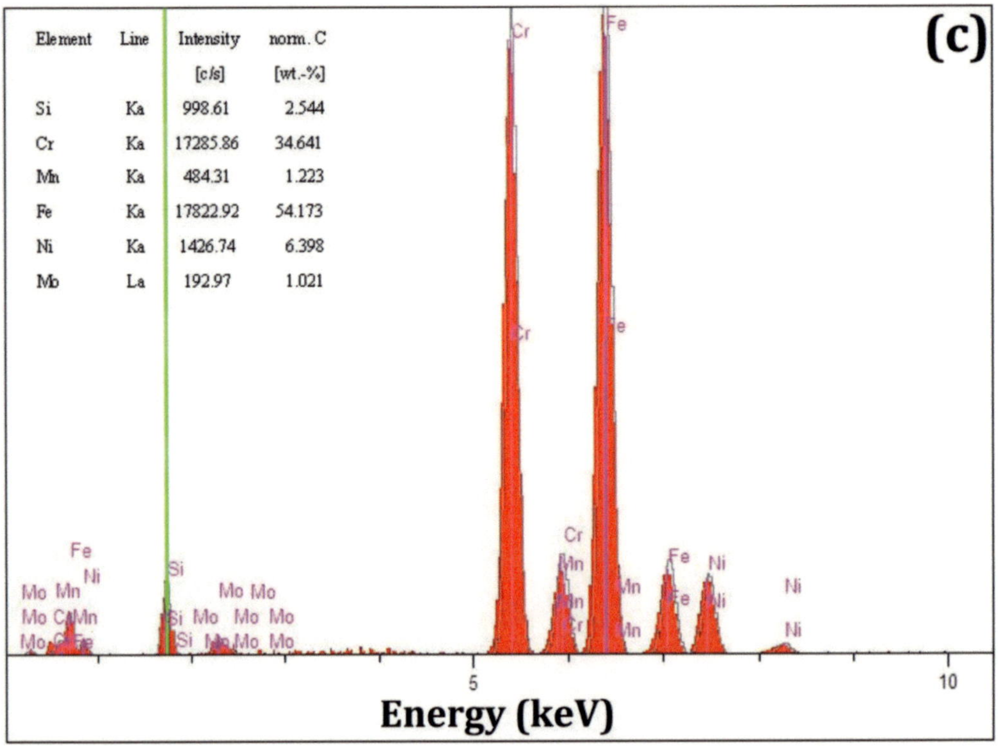

Figure 11. The EDS analysis of: (a) austenite-phase (point 1 in Fig. 11(a)), (b) ferrite-phase (point 2 in Fig. 11(b)), and (c) sigma-phase (point 3 in Fig. 11(c)).

Point	Element (wt.%)		
	Cr	Si	Mo
1 (gamma)	22.125	1.353	0.227
2 (delta)	27.639	2.896	1.341
3 (sigma)	34.641	2.544	1.021

Table 2. Elemental composition of Fig. 12.

Fig. 13 shows the fracture surfaces of the failed sample before homogenization (Fig. 13(a)), the sample after the homogenization heat treatment at 1100 °C for 2 h followed by cooling in furnace (Fig. 13(b)), and the sample after the homogenization heat treatment at 1100 °C for 2 h followed by cooling in air (Fig. 13(c)). The EDS analysis of points 1 and 2 in Figs. 13(a) and 13(b), respectively indicated the presence of sigma-phase in definite crystallographic planes. In fact, the fracture surface of failed sample (Fig. 13(a)) contained a random arrangement of flat surfaces and surfaces with steps which is characteristic of brittle fracture. The fracture surface of the sample after the homogenization heat treatment at 1100 °C for 2 h followed by cooling in furnace (Fig. 13(b)) was a mixture of small dimples and layered surfaces that indicated brittle and ductile fractures in this sample. The EDS analysis of point 3 in Fig. 13(b) showed the presence of chromium in the form of carbide particles. The carbide

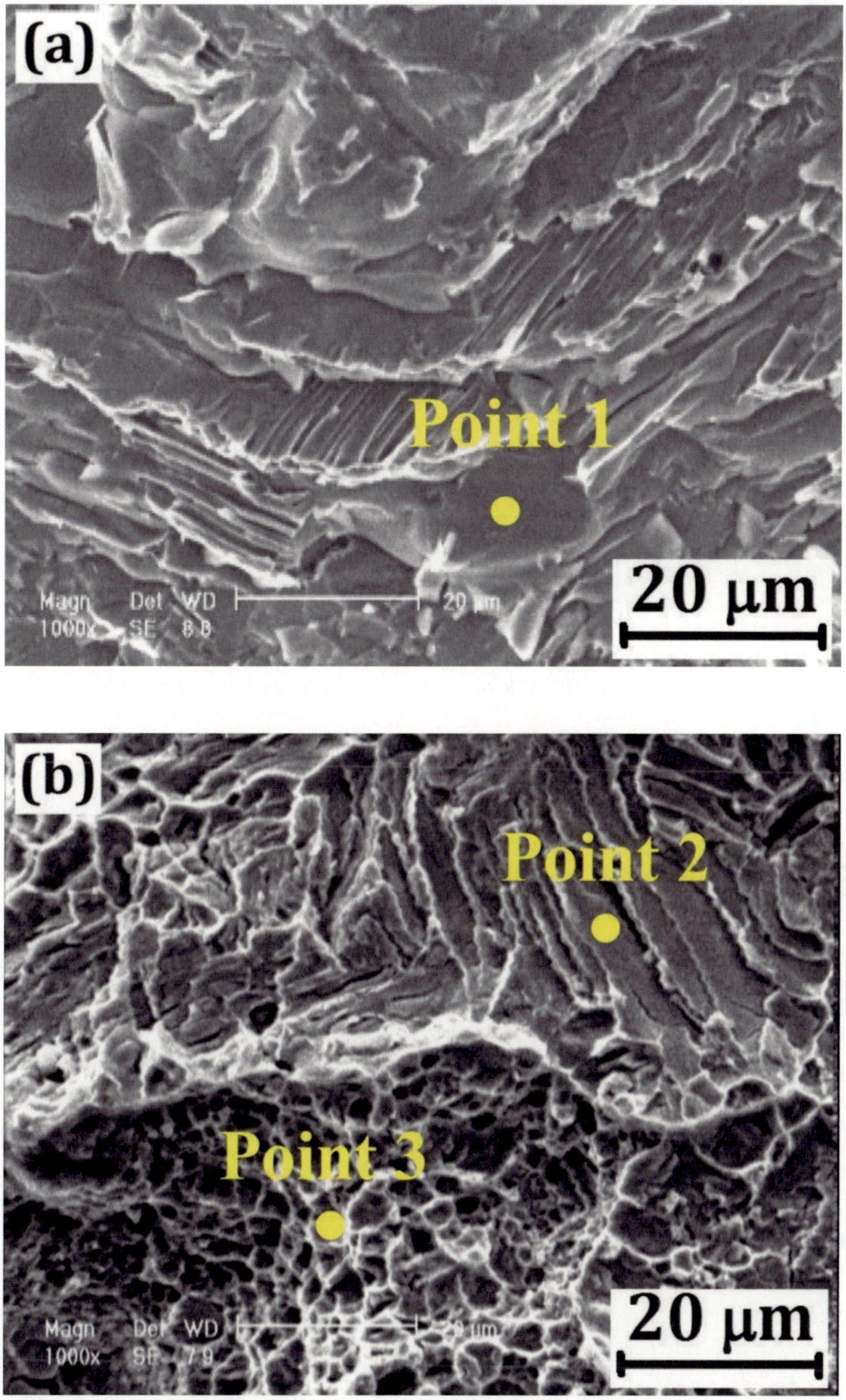
(a)
Point 1
20 μm
Magn Det WD
1000x SE
(b)
Point 2
Point 3
20 μm
Magn Det WD
1000x

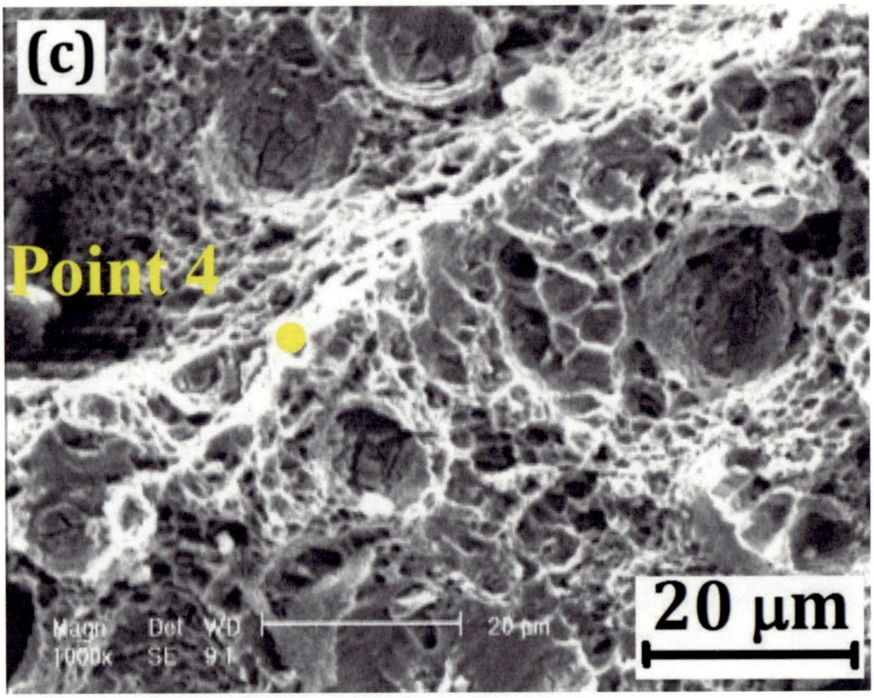

Figure 12. Fracture surfaces of: (a) the failed sample before homogenization, (b) the sample after the homogenization heat treatment at 1100 °C for 2 h followed by cooling in furnace, and (c) followed by cooling in air.

particles existed in the center of small dimples. According to EDS analysis of points 2 and 3, it can be said that due to slow cooling rate in the furnace, all the desirable changes that occur in homogenized structures are reversed and a new formation of the sigma-phase and precipitates in the sample results in a transition from ductile to almost brittle fracture. As can be seen in Fig. 13(c), formation of deep dimples is a main characteristic of ductile fracture. The EDS analysis of the point 4 was similar to the base metal. This figure indicated a considerable decrease in sigma phase and precipitates amounts after homogenization.

Therefore, before roller straightening, the homogenization heat treatment should be performed for dissolution of deleterious phases in austenite as the matrix phase and obtaining good toughness. In many cases, microstructural inhomogeneity can be eliminated by holding the sample at high temperatures for a sufficient time and cooling at an appropriate rate. Otherwise, the possibility of the formation of fine cracks during roller straightening is very high. These cracks can result in roller fracture during the straightening process and/or in the continuous-annealing furnace.

6. Conclusions

The sigma phase tends to precipitate in the regions of high chromium content, such as the chromium carbides in the grain. A relatively small quantity of the σ phase, when it is nearly continuous at a grain boundary, can lead to very early failure of high-temperature parts. It is important to understand that the sigma-phase cannot undergo any significant plastic deformation; instead, it fractures even at relatively low strain levels. This is true at elevated temperatures, but even more so at ambient temperature. Where toughness and ductility are an important part of the system design, the sigma-phase cannot be tolerated. The formation of sigma-phase with dendritic morphology results in decrease in ductility and failure of heat resistant steels. The microstructural inhomogeneity (sigma-phase and carbides) can be eliminated by holding the sample at high temperatures for a sufficient time and cooling at an appropriate rate. The homogenization heat treatment under appropriate condition can led to change in the morphology of sigma phase from dendritic structure to globular structure. This structure leads to remarkable increase in the impact energy and ductility of heat resistant steels.

Author details

Mohammad Hosein Bina
Department of Advanced Materials and New Energy, Iranian Research Organization for Science and Technology, Tehran, Iran

7. References

[1] ASM Handbook, Volume 1, Properties and Selection: Irons, Steels, and High-Performance Alloys, ASM International, Materials Park, Ohio, 2005.

[2] Lamb, S., Practical Handbook of Stainless Steels and Nickel Alloys, ASM International, Materials Park, Ohio, 1999.

[3] ASM Handbook, Volume 15, Casting, ASM International, Materials Park, Ohio, 2005.

[4] Shi, S. and Lippold, J.C., "Microstructure Evolution During Service Exposure of Two Cast, Heat-Resisting Stainless Steels - Hp Nb Modified and 20-32nb", Materials Characterization, Vol. 59, No. 8, pp. 1029–1040, 2008.

[5] Prager, M. and Svoboda, J., Cast High Alloy Mettalurgy, Steel Casting Metallurgy, Steel Founder's Society of America, Rocky River, OH, 1984.

[6] W. Trietschke and G. Tammnann, The Alloys of Iron and Chromium. Zh. Anorg. Chem., Vol 55, 1907, p 402–411.

[7] E.C. Bain and W.E. Griffiths, An Introduction to the Iron-Chromium-Nickel Alloys, Trans. AIME, Vol 75, 1927, p 166–213.

[8] P. Chevenard, Experimental Investigations of Iron, Nickel, and Chromium Alloys, Trav. Mem., Bur. Int. Poids et Mesures, Vol 17, 1927, p 90.

[9] F. Wever and W. Jellinghaus, The Two-Component System: Iron-Chromium, Mitt. Kaiser-Wilhelm Inst., Vol 13, 1931, p 143–147.

[10] E.O. Hall and S.H. Algie, The Sigma Phase, Metall. Rev., Vol 11, 1966, p 61–88.

[11] ASM Handbook, Volume 11, Failure Analysis and Prevention, ASM International, Materials Park, Ohio, 2002.

[12] Bina, M.H., Dini, G., Vaghefi, S.M.M., Saatchi, A., Raeissi, K. and Navabi, M., "Application of Homogenization Heat Treatments to Improve Continuous-Annealing Furnace Roller Fractures", Engineering Failure Analysis, Vol. 16, No. 5, pp. 1720–1726, 2009.

[13] Dini, G., Bina, M.H., Vaghefi, S.M.M., Raeissi, K., Safaei-Rad, M. and Navabi, M., "Failure of a Continuous-Annealing Furnace Roller at Mobarakeh Steel Company", Engineering Failure Analysis, Vol. 15, No. 7, pp. 856–862, 2008.

[14] A.J. Lena and W.E. Curry, The Effect of Cold Work and Recrystallization on the Formation of the Sigma Phase in Highly Stable Austenitic Stainless Steels, Trans. ASM, Vol 47, 1955, p 193–210.

[15] ASM Handbook, Volume 19, Fatigue and Fracture, ASM International, Materials Park, Ohio, 2005.

[16] ASM Handbook, Volume 12, Fractography, ASM International, Materials Park, Ohio, 2005.

[17] Lin, D.Y., Liu, G.L., Chang, T.C. and Hsieh, H.C., "Microstructure Development in 24Cr-14Ni-2Mn Stainless Steel after Aging under Various Nitrogen/Air Ratios", Journal of Alloys and Compounds, Vol. 377, No. 1-2, pp. 150-154, 2004.

[18] Tseng, C.C., Shen, Y., Thompson, S.W., Mataya, M.C. and Krauss, G., "Fracture and the Formation of Sigma Phase, M23C6, and Austenite from Delta-Ferrite in an AISI 304L Stainless Steel", Metallurgical and Materials Transactions A, Vol. 25, No. 6, pp. 1147–1158, 1994.

[19] Martins, M. and Casteletti, L.C., "Sigma Phase Morphologies in Cast and Aged Super Duplex Stainless Steel", Materials Characterization, Vol. 60, No. 8, pp. 792–795, 2009.

[20] Hsieh, C.C., Lin, D.Y. and Wu, W., "Precipitation Behavior of σ Phase in 19Cr-9Ni-2Mn and 18Cr-0.75Si Stainless Steels Hot-Rolled at 800 °C with Various Reduction Ratios", Materials Science and Engineering A, Vol. 467, No. 1-2, pp. 181–189, 2007.

[21] Gill, T.P.S., Vijayalkshmi, M., Rodriguez, P. and Padmanabhan, K.A., "On Microstructure-Property Correlation of Thermally Aged Type 316L Stainless Steel Weld Metal", Metallurgical Transactions A, Vol. 20, No. 6, pp. 1115–1124, 1989.

[22] Hsieh, C.C., Lin, D.Y. and Chang, T.C., "Microstructural Evolution During the δ/σ/γ Phase Transformation of the Sus 309L Si Stainless Steel after Aging under Various Nitrogen Atmospheric Ratios", Materials Science and Engineering A, Vol. 475, No. 1-2, pp. 128–135, 2008.

[23] G.N. Emanuel, Sigma Phase and Other Effects of Prolonged Heating at Elevated Temperatures on 25 Per Cent Chromium-20 Per Cent Nickel Steel, in Symposium on the Nature, Occurrence, and Effects of Sigma Phase, STP 110, American Society of Testing and Materials, 1951, p 82–99.

[24] G. Aggen et al., Microstructures Versus Properties of 29-4 Ferritic Stainless Steel, in MiCon 78: Optimization of Processing, Properties, and Service Performance Through Microstructural Control, STP 672, American Society for Testing and Materials, 1979, p 334–366.

Chapter 6

Multiconvolutional Approach to Treat the Main Probability Distribution Functions Used to Estimate the Measurement Uncertainties of Metallurgical Tests

Ion Pencea

Additional information is available at the end of the chapter

http://dx.doi.org/10.5772/50511

1. Introduction

The quality of a metallic product is validated by tests results. Thus, the product quality is depicted by the quality of the testing results that furthermore depends on several factors such as: the incomplete knowledge about measurand, the adequacy of the testing method in relation to measurand, the equipment adequacy for method, the human factor; the statistical inference, etc. The influence factors of a measurement process, whether known or unknown, may alter the result of a measurement in an unpredictable way. Thus, a test result bears an intrinsic doubt about its closeness to the conventional true value of the measurand, fact which is commonly perceived as uncertainty about the test result. One of the most important tasks for the experimentalist is to specify an interval about the estimated value of the measurand ($\bar{x}$), $[\bar{x} - U; \bar{x} + U]$ in which the true (conventional) value (μ) could be found with specified confidence level, i.e. the probability (p) that $\mu \in [\bar{x} - U; \bar{x} + U]$. The practice exigency requires $p \geq 95\%$[1, 2, 3]. The EN ISO 17025 [1] stipulates that the quality of a numeric test result is quantified by the expanded uncertainty $U(p\%)$, where p is the level of confidence ($p \geq 95\%$). An alternative specification of U is its level of significance expressed as $1-p$. In order to obtain a higher quality of the test result, the experimentalist should perform a set of at least 30 repeated measurements [2, 4]. In the field of metallurgy this is quite impossible for technical and economical reasons. Therefore, the quality of the test result should be guaranteed by advanced knowledge about the testing method and by other means such as: equipment etalonation, Certified Reference Materials (CRMs) usage and, last but not least, correct uncertainty estimation based on proper knowledge about the probabilistic behavior of the compound random variable derived from a set of repeated measurements (arithmetic mean, standard deviation, etc.). This chapter is intended to provide the knowledge for a better statistical evaluation of the outcomes of the metallurgical tests, based on proper selection of the probability density function mainly for compound variable such as arithmetic sample mean, standard deviation, t-variable etc. The reader will find in this chapter the derivations of the Gauss, Student, Fisher-Snedecor distributions and also several compound ones. The derivations are intended to provide the information necessary to select the appropriate specific distribution for a measurand. Furthermore, the chapter addresses the uncertainty estimation based on multiconvolutional approach of the measurand by presenting a case study of Rockwell C hardness test, that reveals the superiority of the statistical inference based on the approach proposed in this chapter.

2. Probability density functions for experimental data analysis

2.1. Elements of Kolmagorov's theory of probability

According to Kolmogorov's theory [4, 5], the behavior of a random experiment or phenomenon can be mathematically modeled in the frame of class theory using the sample space, the event class and the probability function [5, 6]. Kolmagorov's theory addresses experiments, phenomena or classes of entities having likelihood behavior that can be repeated under the same condition as many times as needed. The testing of the occurrence of an event will be considered generically as being an experiment or probe. The sample space or sample universe (*E*) is the entire class of outcomes e_i, $i=\overline{1,n}$, of an experiment that are mutually exclusive events, respectively

$$E = \{ei\}\, i = \overline{1,n} \text{ where } e_i \cap e_j = \phi\ \forall\, i \neq j\,, i,j = \overline{1,n} \tag{1}$$

An event *A* is a part of the *E*, i.e. $A \in \mathcal{P}$ (E). The probability of an event is a function defined on $\mathcal{P}$ *(E)* i.e. P: $\mathcal{P}(E) \rightarrow [0;1]$, which satisfies the following axioms:

$$P(A) \geq o\,; \forall\, a \in \mathcal{P}(\Omega) \tag{2}$$

$$P(E) = 1 \tag{3}$$

$$P(A \cup B) = P(A) + P(B)\,; if\ A \cap B = \phi \tag{4}$$

2.2. Discrete and continuous random variables

A discrete random variable is associated to an experiment that gives finite or countable elementary outcomes, having well defined probabilities. The sum of the probabilities of the discrete sample space must be one. To a finite set of outcomes of an experiment, a discrete random variable X is assigned, which is represented as:

$$X \begin{pmatrix} x_1\, x_2 \, \ldots \ldots \ldots & x_n \\ p_1\, p_2 \, \ldots \ldots \ldots & p_n \end{pmatrix} \tag{5}$$

For a countable set of outcomes, a discrete random variable *X* is expressed as:

$$X\begin{pmatrix} x_1 & x_2 \dots & x_k \dots \dots \\ p_1 & p_2 \dots\dots & p_k \dots \dots \end{pmatrix} \tag{6}$$

The relationships in (5) and (6) are called the probability distribution functions (*pdf*) of the discrete variable. As it is well known, there are experiments given numeric continuous outcomes. To such experiments, continuous variables are associated. For a continuous random variable X, the probability assigned to any specific value is zero, whereas the probability that X takes values in an interval [*a*, *b*] is positive. The probability that a continuous random variable X takes values in the *[a, b]* interval is expressed as $P(a < X < b)$. The probability that a continuous random variable X is less than or equal with a value *x* is

$$F_X(x) = P\,(X \le x) \tag{7}$$

which is called cumulative distribution function (*cdf*).

Fx(x) should be a continuous and derivable function to fulfill the condition that for any infinitesimal interval [*x*, *x*+*dx*] one can estimate the probability that $X \in [x, x + dx]$ as:

$$\mathrm{dP}((x \le X < x + dx\,) = F(x + dx) - F(x) = \frac{F(x+dx) - F(x)}{dx} \cdot dx = \frac{dF}{dx} \cdot dx = p\,(x)dx \tag{8}$$

where $p(x) = \frac{d\,F_x(x)}{dx}$ is the density distribution function of X. As it is evident, $F_X(x) = 0$ for $x \le a$ while $F_X(x) = 1$ for $x \ge b$.

2.3. Independent and conditional events

2.3.1. Conditional probabilities

Let *E* be a discrete sample space, containing *n* elementary events of an experience with probabilistic outcomes. Let be two events *A* and *B* of *E* that contain *k* , respectively *l* elementary events so that $A \cap B$ contains *m*. Assuming a trial is done and *B* occurs, then the probability that *A* occurs simultaneously is the ratio of favorable outcomes for *A* contained in *B* and of the possible outcomes of *B*. Thus, the probability of the event *A*, knowing that the compatible event *B* occurred is named conditional probability of *A* given *B* and denoted as $P_B(A)$ *or* $P(A \cap B)$. In the above example the $P_B(A)$ is:

$$P_B(A) = \frac{m}{l} = \frac{\frac{m}{n}}{\frac{l}{n}} = \frac{P(A \cap B)}{P(B)} \tag{9}$$

The probability of *B* given *A* is:

$$P_A(B) = \frac{m}{k} = \frac{\left(\frac{m}{n}\right)}{\left(\frac{k}{n}\right)} = \frac{P(A \cap B)}{P(A)} \tag{10}$$

From Eqs.(9) and (10) it can be derived that

$$P(A \cap B) = P(A) \cdot P_A(A \cap B) = P(B) \cdot P_B(A) \tag{11}$$

Event A is independent of B if the conditional probability of A given B is the same as the unconditional probability of A, $P(A)$ e.g $P_B(A) = P(A)$. According to Eg.(11) the probability of two independent event of E , let say A and B, is:

$$P(A \cap B) = P(A) \cdot P(B) \quad (12)$$

2.3.2. *Pairwise and global independence*

If the event $A_i, i = \overline{1,3}$ are such that any of pairs are exclusive e.g. $A_i \cap A_j = \emptyset, \forall i \neq j$ then the events $A_i, i = \overline{1,3}$ may be not totally exclusive i.e. $A_1 \cap A_2 \cap A_3 \neq \emptyset$. The classical proof of 3 events pairwise independent but not totally independent was given by S. Berstain [5, 6]. The events $A_i \subset E, i = \overline{1,n}$ are totally independent if for any selection of k events of E, written as $\{A_{S1}, A_{S2}.......A_{Sk}\}$, the following statement is true

$$P(\cap_{i=1}^{k} A_{Si}) = \prod_{i=1}^{k} P(A_{Si}) \quad (13)$$

2.3.3. *Geometric probabilities*

The probability of an event related to the location of a geometric figure placed randomly on a specific planar or spatial domain is called geometric probability. A representative example is that of a disk (D) of radius "r", that is thrown randomly onto a planar domain A (square of edge length a) that includes a sub domain B (square of edge length b) as shown in Figure 1.a. The addressed problem is to estimate the possibility that the disk center falls into the domain B. This is the ration between the area of domain B and the area of A, i.e. $P(C \in B) = (b/a)^2$. If the event consists of $D \cap B$, then the probability is $P(D \cap B) = (b^2+4br+\pi r^2)/a^2$. The examples could be extended to the micro hardness test, e.g. if a field of a steel specimen of area A contains compounds of total area B (Figure 1.b) then the probability that at a random indentation the indenter tip impinges on a compound is $p_B = B/A$. The most well-known example of geometric probability is the Buffon's needle problem [7]. Thus, if a needle of length "$2l$" is dropped randomly on a horizontal surface ruled with parallel lines at a distance $h > 2l$ apart what is the probability that the needle intersects one of the lines?

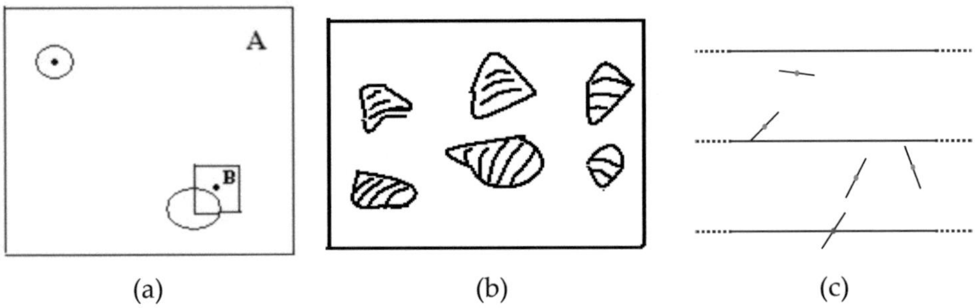

Figure 1. Schematic representation of the sample space for: a) disk thrown; b) micro-indentation; c) Buffon's needle problem

The probability that a needle thrown randomly crosses a line is the sum of all favorable probabilities which, in fact, is the integral:

$$P = \int_0^{\pi} \frac{l\, sin\theta}{h} \cdot p(\theta)d\theta = \frac{2l}{\pi h} \tag{14}$$

The criterion for estimating the correctness of a Buffon's needle experiment is the closeness of $(2l)/(hP)$ to the value of π. The geometrical probabilities are used on a large scale in metallography for grain size and grain shape estimation [7]. Their usage may be extended to micro and nano-indentation tests etc.

2.4. Discrete probability density functions

Discrete probability deals with events that occur in a countable sample space [1-4]. If the discrete sample contains "n" elementary events $\{e_i; i = \overline{1,n}\}$ then an "intrinsic" probability value is assigned to each e_i while for any event X is attributed a probability $f(x)$ which satisfies the following properties:

$$f(X) \epsilon\ [0,1] \text{ for all } X\ \epsilon\ E \tag{15}$$

$$\sum_{X \epsilon E} f(x) = 1 \tag{16}$$

The function $f(ei), i = \overline{1,n}$ that maps an elementar event to the "probability value" in the [0, 1] interval is called the probability mass function, abbreviated as *pmf*. The probability theory does not deal with $f(e_i)$ assessing, but builds methods for probability calculation of every event assuming prior known of $f(e_i)$, $i = \overline{1,n}$. The *pmf* is synonymous with *pdf* therefore *pdf* will be used as the abbreviation for the probability density of X, which is:

$$X\begin{Bmatrix} x_1, x_2, \dots \dots \dots x_n \\ p_1, p_2, \dots \dots \dots p_n \end{Bmatrix} \tag{17}$$

where $x_i = X(e_i)$, $p_i = f(e_i) \equiv f(x_i)$; $i = 1, n$

Since the most used *pdfs* ("*pmfs*") in the field of metallurgy are Poisson and Bernoulli distribution only these distributions will be addressed in this chapter.

2.4.1. Poisson scheme

Let us consider a set of n experiences denoted as: $E = E_1 x E_2\ x \dots . x\ E_n$. Each experience has two elementary events, i.e. the variable X_i attached to the sample space of E_i, $i = \overline{1,n}$ is:

$$X_i \begin{pmatrix} A_i & \overline{A_i} \\ p_i & q_i \end{pmatrix} \tag{18}$$

where A_i is the event of interest of E_i, having the probability "p_i" while $\overline{A_i}$ is the contrary one having the probability $q_i = 1 - p_i$.

The Poisson's scheme is designed to help estimating the probability of occurrence of k expected outcomes when each experiment Ei, $i = \overline{1,n}$, of E is performed once. In this instance, assuming that "*k*" expected events occurred then one can renumber the events starting with these "*k*" expected and, next, with those (n-k) unexpected as follows:

$$\{A_{j1}, A_{j2,} \dots \dots \dots . A_{jk}, \bar{A}_{jk+1}, \bar{A}_{jk+2} \dots \dots \dots \bar{A}_{jn}\} \tag{19}$$

The probability of the event $E_j = A_{j1} \cap A_{j2} \cap \dots \dots \cap A_{j+n}$ is the product of the individual A_{jl} events:

$$P(E_j) = p_{j1}, p_{j2}, \dots q_{jk+1} \dots . q_{jn} \tag{20}$$

The event $E_n(k)$ is realized for any combination of k A_i, $i = \overline{1,n}$, events i.e. for C_n^k events. Thus to calculate the $P(E_n(K)) \equiv P_n(K)$ one must sum all different products consisting of "*k*" terms of p_i, $i = \overline{1,n}$ multiplied by the rest of q_i probabilities. This way of calculation is identical with the case of calculating the coefficient of X^k of the polynomial:

$$P_n(X) = (p_1X + q_1)(p_2X + q_2) \dots . (p_nX + q_n) \equiv \prod_{i=1}^{n}(p_iX + q_i) \tag{21}$$

This approach of calculations of $P_n(k)$ as the coefficient of X^k of $P_n(x)$ is known as Poisson's scheme.

2.4.2. Bernoulli or binominal distribution

The Bernoulli distribution addresses the case of an array of "*n*" identical experiences, i.e. the particular case of Poisson scheme when $p_i = p$ and $q_i = q$, $i = \overline{1,n}$. In this case, the probability of occurrence of "*k*" events from "*n*" trials is:

$$P_n(k) = C_n^k p^k q^{n-k} = \frac{n!}{k!(n-k)!} p^k q^{n-k} \tag{22}$$

where n!=1*2*3*...*n

The mean value of "*k*", denoted as μ, could be calculated as:

$$M(k) = \sum_{k=0}^{n} kC_n^k p^k q^{n-k} = \sum_{k=1}^{n} nC_{n-1}^{k-1} p^k q^{n-k} = \sum_{j=0}^{n-1} npC_{k-1}^{n-1} p^j q^{n-1-j} = np \tag{23}$$

The term dispersion is many times used instead of variance.

The variance of the Bernoulli distribution is:

$$V(k) = \frac{1}{n}\sum_{k=1}^{n}(k-\mu)^2 \sum \cdot p_n(k) = \frac{1}{n}(\sum_{k=1}^{n} k^2 P(k)) - \mu^2 = \mu^2 p^2 q^2 \tag{24}$$

2.4.3. Poisson distribution

Poisson distribution is a particular case of the Bernoulli distribution when the number of events tested is very large, but the probability of the experimental outcome is close to zero i.e. it is the distribution of rare events. In this instance, the mean $\mu = n \cdot p$ is considered a constant quantity that characterizes the distribution as will be shown forwards. According to the Bernoulli distribution the probability of k events realization in a series of $n \to \infty$ is:

$$P_n(k) = C_n^k p^k q^{n-k} = \frac{n(n-1)..(n-k+1}{k!} p^k (1-p)^{n-k} = \prod_{i=0}^{k-1} \left(1 - \frac{i}{n}\right) * ([\left(1 - \frac{\mu}{n}\right)^{-\frac{n}{\mu}}]^{-\mu\frac{(n-k)}{n}} \quad (25)$$

The limit of $P_n(k)$ for $\rightarrow +\infty$, denoted as *P(k),* is:

$$P(k) = \lim_{n \to \infty} (P_n(k)) = \frac{\mu^k \cdot e^{-\mu}}{k!} \quad (26)$$

The dispersion of the Poisson distribution is $D^2 = \sqrt{\mu}$.

The Poisson distribution is often assigned to the quantum particle counting statistics because the standard deviation (SD) is simply expressed as square root of the mean number $<n>$, where $<n>$ is the mean of a set of measurements. Many times the SD is estimated as *sqrt(n)* using a single test result. But this approach is many times inappropriate to the real situation because the detection probability of a quatum particle is not close to zero.

There are other interesting discrete *pdfs* for metallurgist as: hypergeometric distribution, negative binomial distribution, multinomial distribution, [1-5] but it is beyond the scope of this chapter.

2.5. Continuous probability density function

2.5.1. Introduction

The variable X is called a continuous random variable if its values are real numbers as the outcomes of a large class of measurements dealing with continuous measurand such as temperature of a furnace, grain size etc. The sample space for a continuous variable X may be an interval *[a, b]* on the real number axes or even entire $\mathbb{R}$ space. The probability thar X takes a value in the interval $[x_1, x_2]$, denoted as $P(x_1<X<x_2)$, is directly proportional with the interval length $|x_2-x_1|$ and depends on the intrinsic nature of the experiment to which X was assigned. The local specificity of X is assessed by $P(x<X<x+dx)$, where *dx* is of infinitesimal length. The probability that $X<x$ is $P(X<x)= F(x)$, which is called cumulative distribution function, abbreviated *cdf*. By definition, the *cdf* must be derivable because $P(x<X<x+dx)= F(x+dx)-F(x)$ and for any *dx* the $(F(x+dx)-F(x))/dx$ should be finite and continuous i.e.

$$\lim_{dx\to 0}(\frac{F(x+dx)- F(x)}{dx}) = p(x) \quad (27)$$

where *p(x)* is the probability density function (*pdf*) over the range $a \le X \le b$.

The main statistical parameter assigned to a variable X on the base of its *pdf* is: mean, median, mode, sample variance or dispersion, standard deviation and central moments. The mean of X variable is:

$$\mu \equiv M(X) = \int_a^b x \cdot p(x)dx \quad (28)$$

where *[a, b]* is the domain of X variable.

The median value of X (x_m) is the value which divides the number of outcomes into two equal parts i.e. $F(x_m)$=0.5. The mode of a *pdf* is the value x_0 for which $p(x_0)$ reaches its maximum i.e. $dp(x_0)/dx=0$

The variance V of X is expressed as:

$$V_X = \int_a^b (x-\mu)^2 p(x)dx \tag{29}$$

The standard deviation is defined as:

$$SD_X = sqrt(V_{X)} \tag{30}$$

SD is a measure of the spreading of X values about μ. A set of measurement results are even more centered around μ as their dispersion is smaller.The central moment of the order r is defined as:

$$M_r(X) = \int_a^b (x-\mu)^r p(x)dx \tag{31}$$

where r is a natural number and usually $r>2$.

The central moments are used for assessing the skewness and kustoisis of the *pdfs*. [1, 4]

2.5.2. Continuous uniform probability distribution function

A random continuous variable X has a uniform *pdf* if it takes values with the same probability in an interval *[a, b]*. The *pdf* of an uniform variable, abbreviated as *updf*, is given as:

$$p(x) = \begin{cases} \frac{1}{b-a}\ ;\ x \in [a,b] \\ 0;\ x \notin [a,b] \end{cases} \tag{32}$$

The *cmd* of X is:

$$\int_a^x f_x(t) = F_X\,(x) = \begin{cases} 0;\ x \le a \\ \frac{x-a}{b-a}\ ; x \epsilon\ [a,b] \\ 1\ ;\ x \ge b \end{cases} \tag{33}$$

Usually, a *updf* is transformed in a standard uniform distribution by the coordinate transformation:

$$z = \left[x - \frac{(a+b)}{2}\right] \cdot \frac{2}{b-a} \in [-1,1] \tag{34}$$

which leads to:

$$f(z) = \begin{cases} \frac{1}{2}\ ;\ x \in [-1,+1] \\ 0;\ x \notin [-1,+1] \end{cases} \tag{35}$$

The *cdf* of a standard updf is:

$$f(z) = \begin{cases} 0\ ;\ x < -1 \\ \frac{1}{2}\ (1+x)\ ;\ x \in [-1,+1] \\ 1\ ;\ x > 1 \end{cases} \tag{36}$$

The graphs of the standard *updf*, *f(x)*, and of its *cdf*, *F(x)*, are shown in Figure 2.

The main parameters of a *updf* are: $\mu = \frac{a+b}{?}; SD = (b-a)/\sqrt{3}; \mu_{2k+1} = 0$ while for a standard *updf* are $\mu=0$, $SD=\sqrt{3}/3$. In testing practice, a *updf* is assigned to an experimental measurand when there is no information or experimental evidences that its values have a clustering tendency. This is the case of a measure device having specified only the tolerance range as for the pyrometer i.e. ±5 ºC.

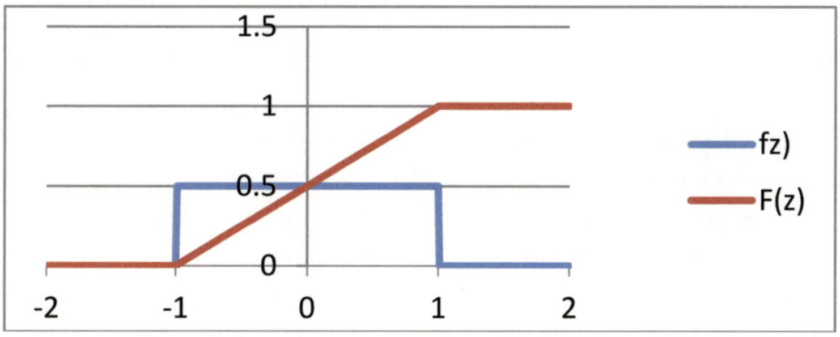

Figure 2. The graphs of the standard *updf, f(x)*, and of its *cdf, F(x)*

2.5.3. *Trapezoidal probability distribution function*

A trapezoidal *pdf* is ascribed to a continuous random measurand if the distribution of its values around the mean is likely to be uniform while at the extremities of the interval the frequency of its occurrence vanishes linearly to zero as is shown in Figure 3.

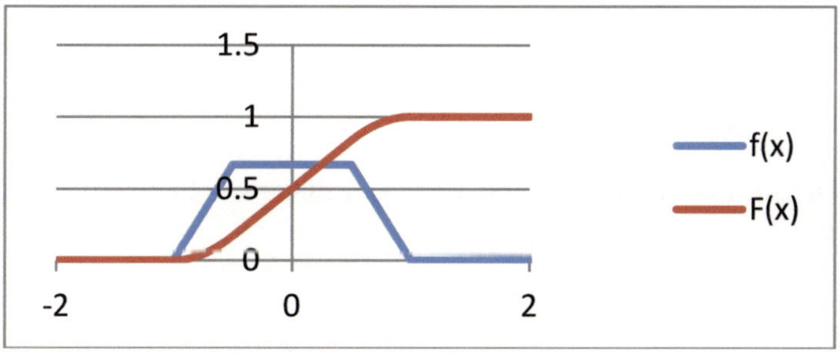

Figure 3. The graphs of trapezoidal symmetric *pdf, f(x)*, and of its *cdf, F(x)*.

The length of the larger base of the trapeze is usually denoted as $2a$ while the lesser one is $2b=2a\beta$. The height of the trapeze (h) is determined by the normalization condition i.e. the area between $f(x)$ and Ox axis is 1. Thus, for a trapezoidal *pdf* $h = [a(1+\beta)]^{-1}$. In this instance, the isosceles trapezoidal *pdf* is expressed as:

$$f(X)\begin{cases} \frac{x-\mu+a}{a^2(1-\beta^2)};\ x \in [\mu-a; \mu-\beta a] \\ \frac{1}{a(1+\beta)};\ x \in [\mu-\beta a; \mu+\beta a] \\ \frac{x-\mu-\beta a}{a^2(1-\beta^2)};\ x \in [\mu+\beta a; \mu+a] \\ 0;\ otherwise \end{cases} \quad (37)$$

The $F(x)$ of a trapeze *pdf* is obtained by integrating $f(x)$ over the interval [μ-*a; x*].

The variance of *f(x)* given by Eq.(37) is:

$$V = a^2\left(1+\beta^2\right)/6 \tag{38}$$

The trapezoidal distribution could be considered as the general form for the class of distribution whose graphs are made of linear branches. Thus, for β=1 the trapezoidal *pdf* degenerates into a uniform distribution and for β=0 into a triangular one. Therefore, the triangular distribution will be considered as a particular case of trapezoidal which has the following parameters: μ, and $V=a^2/6$. The triangular distribution is appropriate for the measurand whose values cluster around the mean value. The triangular distribution with the width 2*a* may be considered as a twofold convolution of uniform distribution of identical length *a*. The same, the trapezoidal distribution can be seen as a convolution of two different uniform distributions. The triangular *pdf* is mostly used for uncertainty estimation of type B given by an instrument whose tolerance limits are specified.

2.6. The normal or Gaussian probability distribution function

Herein will be presented a derivative of normal *pdf* to emphasize the circumstances of its application The normal or Gaussian *pdf* was formulated by F. Gauss in 1809 and since then it became the most prominent *pdf* encountered in testing practice. For example, the result error in a test is usually assumed to follow a normal distribution. The Gaussian function is defined on the entire axis of real number as [4-6]:

$$f(x;\mu,\sigma^2) = \frac{1}{\sigma\sqrt{2\pi}}\, e^{-\frac{(x-\mu)^2}{2\sigma^2}} \tag{39}$$

where μ is the mean and σ^2 is the variance of the continuous random variable X.

The Gaussian distribution with μ=0 and σ^2=1 is called the standard normal distribution, denoted by $\Phi(x)$ or *N(0,1)* which is expressed as:

$$\phi(x) = \frac{1}{\sqrt{2\pi}}\, e^{\frac{-x^2}{2}} \tag{40}$$

$N(\mu,\sigma^2)$ can be expressed as:

$$f(x;\mu,\sigma^2) = \frac{1}{\sigma}\Phi\left(\frac{x-\mu}{\sigma}\right) \tag{41}$$

The function *cdf* of the standard normal distribution is:

$$\phi(x) = \frac{1}{\sqrt{2\pi}}\int_{-\infty}^{x} e^{-\frac{t^2}{2}}\, dt \tag{42}$$

The integral (40) cannot be expressed in terms of elementary functions but as error function as:

$$\Phi(x) = \frac{1}{2}\left[1+\operatorname{erf}\left(\frac{x}{\sqrt{2}}\right)\right], x \in R \tag{43}$$

The *cdf* of $N(\mu,\sigma^2)$, $F(x;\mu,\sigma^2)$, can be expressed as:

$$F(x;\ \mu, \sigma^2) = \Phi(\tfrac{x-\mu}{\sigma}) \tag{44}$$

The normally distributed variable has a symmetric distribution about μ, which is in the same time the median and the mode. The probability that $X \sim N(\mu,\sigma^2)$ takes values far from μ (i.e. more than a few standard deviations) drops off extremely rapidly.

From the perspective of the test result analysis, it is important to derive the $f(x;\mu,\sigma^2)$. An easy and meaningful approach to derive $f(x;\mu,\sigma^2)$ is based on binomial *pdf* with p not close to zero and for a very large n. In such a case, $P_n(k)$ has a maximum for $\tilde{k} = np$ and drops off very rapidly as k departs from $\tilde{k}$. When n is very large, $P_n(k)$ varies smoothly with k and practically k can be considered a continuous variable. Because $P_n(k)$ takes significant values only in the vicinity of k, then its values will be well approximated by formally constructed Taylor series for $\ln(P_n(k))$, around $\tilde{k}$ as follows:

$$ln[P_n(k)] \cong ln[P_n(\tilde{k})] + \frac{(k-\tilde{k})}{1!}\frac{d(\ln(P_n(k))}{dk}\Big|\tilde{k} + \frac{(k-\tilde{k})}{2!}\frac{d^2\, lnP_n(k)}{dk^2}\Big|\,\tilde{k} + \frac{(k-\tilde{k})}{3!}\frac{d^3 lnP_n(k)}{dk^3}|\tilde{k} + \cdots \tag{45}$$

The second term of the right side of the equation is null because $\frac{dP_n(\tilde{k})}{dk} = 0$.

For deriving the $\ln(P_n(k))$ it is assumed that the Stirling's first approximation is valid i.e.

$$ln\frac{n!}{k!(n-k)} \cong n \cdot \ln(n) - - k\ln(k) - (n-k)\ln(n-k) \tag{46}$$

Based on Eq.(46) the derivative of $lnP_n(k)$ of the r order, $r \geq 2$, was deduced as:

$$\frac{d^r ln(P_n(k))}{dk^r} = (-1)^{r-1}.\frac{(r-2)!}{k^{r-1}} + \frac{(r-2)!}{(n-k)^{r-1}} \tag{47}$$

Thus, the terms containing derivative of the order $r > 2$ could be neglected in the Taylor's expression of $P_n(k)$. With these considerations and taking into account that $\frac{d^2 lnP_n(\tilde{k})}{dk^2} = -\frac{1}{npq}$ then Eq.(45) can be written as:

$$P_n(k) \cong P_n(\tilde{k})e^{-\frac{(k-np)^2}{2npq}} = P_n(\tilde{k})e^{-\frac{(k-\mu)^2}{2\sigma^2}} \tag{48}$$

where μ is the mean and σ^2 is the variance of binomial *pdf*

The next assumption based on n sufficient larger is to consider k as a continuous variable X.

The X variable may be extended on R based on exponential decreasing of the probability that X takes values far from μ. In this instance, Eq.(48) can be written as:

$$f(x;\mu,\sigma^2) = \mathrm{C}.e^{-\frac{(x-\mu)^2}{2\sigma^2}} = \frac{1}{\sigma\sqrt{2\pi}}e^{-\frac{(x-\mu)^2}{2}} \tag{49}$$

where C is a constant determined from the normalized condition.

Deduction mathematical expression of the Gauss distribution, $f(x;\mu,\sigma^2)$, on the base of the binomial *pdf* clearly shows that the Gaussian *pdf* is valid for a very large number of experiments. Therefore, by analogy, it is a matter of evidence that Gaussian *pdf* addresses experiments having a large number of influence factors that give rise to random, unbiased errors. Usually, an uncertainty budget comprises a number of influence factors less than 20. Thus, at the first glance, it results that a normal *pdf* is not sufficiently justified to be applied in such a case. On the other hand, if each influence factor has its own influence factors so that the number of the overall contributors to the uncertainty of the measurand exceeds 30, then assigning a normal *pdf* to the measurand is justified.

2.7. Continuous probability distribution functions used in metallurgy practice

In principle, any continuous function defined on an interval $[a, b] \subset R$ can be used as a *pdf* on condition that:

$$\int_a^b f(x)dx = 1 \tag{50}$$

For metallurgists, the most useful *pdfs* other than the normal one are the log-normal, Weibull, Cauchy (Cauchy-Lorentz) and exponential *pdf* [6, 7]. The log-normal *pdf* is used mainly for grain-size data analysis [6]. The mathematical expression of log-normal distribution is derived from normal one by substituting *ln(x)* for *x* as follows:

$$f(x) = \frac{1}{\sqrt{2\pi} ln\sigma_g} \cdot \frac{1}{x} \cdot exp[-\frac{(lnx - ln\mu_g)^2}{2(ln\sigma_g)^2}] \tag{51}$$

where μ_g is the "geometric mean" and σ_g is the "geometric standard deviation".

The log-normal *pdf* is proven by empirical facts. Thus, intended used of a log-normal *pdf* for a specific sample remains at the latitude of the experimentalist.

In the field of metallurgy, the Weibull *pdf* is used mostly for failure rate or hazard rate estimation. The *pdf* of a Weibul random variable X is: [5, 8]

$$f(x; \lambda; k) = \begin{cases} \frac{k}{\lambda} \cdot \frac{x}{\lambda} \cdot e^{-(\frac{x}{\lambda})^k}; \ x \geq 0 \\ 0; \ x < 0 \end{cases} \tag{52}$$

where $k > 0$ is the shape parameter and $\lambda > 0$ is the scale parameter of the distribution.

The Cauchy (Cauchy-Lorentz) function is used successfully to describe the X-ray diffraction or spectral lines profiles which are subjected to homogeneous broadening. The Cauchy function is many times used as a *pdf* of X variable defined on R, as follows [5, 6, 8]:

$$f(x, x_0, \gamma) = \frac{1}{\pi} \left[\frac{\gamma}{\gamma^2 + (x - x_0)^2} \right] \tag{53}$$

where x_o is the peak location and γ is the half-width at half-maximum.

As for normal *pdf, f(x; 0, 1)* is called the standard Cauchy distribution whose *pdf* is:

$$f(x; 0; 1) = \frac{1}{\pi(1 + x^2)} \tag{54}$$

The exponential distribution is fitted to describe the random behavior of a process in which events occur independently at a constant average rate. The *pdf* of the exponential distribution is:

$$f(x;\lambda) = \begin{cases} \lambda e^{-\lambda x};\ x \geq 0 \\ 0;\ x < 0 \end{cases} \tag{55}$$

where $\lambda>0$ is the rate parameter.

The mean of an exponential *pdf* is $\mu = \frac{1}{\lambda}$ while its standard deviation is $\sigma = \frac{\sqrt{2}}{\lambda} = \sqrt{2}\mu$.

The exponential, Cauchy, log-normal and Weibull *pdfs* were presented very shortly for the sake of the chapter completeness but normal and uniform distributions will be used extensively in the next subchapters.

3. The probability distribution function of the compound variables

A random variable Y is defined as a function of other variables X_i, $i=\overline{1,n}$, denoted as $Y=f(X_i)$. In this chapter only the functions met in testing practice are considered and expressed as: $Y=aX+b$, $Y=(X_1+\ldots X_n)/n$; $Y=X^2$ and $Y=\sqrt{X}$; $Y=X_1/X_2$ and $Y=X^2_1/X^2_2$.

3.1. The probability distribution function of the variable $Y=aX+b$

The simplest case of a compound variable is that of the variable $Y=aX+b$ where a and b are two positive real numbers. Assuming that the *pdf* of X is defined on ℝ then the probability probability $P_Y(Y \leq y = aX + b)$ is equal with $P_X(X\leq x)$. In this instance, the *cdfs* of Y and X for $Y = aX+b$ have the same value i.e.

$$F_Y(y) = \int_{-\infty}^{y} f_Y(u)du = F_X(x) = \int_{-\infty}^{x} f_X(v)dv \tag{56}$$

Substituting the variable $t=av+b$ for v in Eq.(56)

$$\int_{-\infty}^{y} f_y(u)du = \int_{\infty}^{Y} \frac{1}{a} f_x\left(\frac{t-b}{a}\right) dt \tag{57}$$

Accordingly,

$$f_y(y) = \frac{1}{a} f_X(\frac{y-b}{a}) \tag{58}$$

3.2. The probability density function of linear compound variables

Consider two random variables X_1 and X_2: ℝ → ℝ. If a variable X is defined on the intervals, it can be considered defined on ℝ because their *pdfs* can be extended to ℝ as follows:

$$f_{X_1}(x) = \begin{cases} f_{X_1};x \in [a,b] \\ 0;x \notin [a,b] \end{cases} \tag{59}$$

The extension to ℝ for $f_{X_1}(x)$ and $f_{X_2}(x)$ is necessary in order to make possible their convolution. The *cdf* of the variable Y is:

$$F_Y(y) = \mathrm{P}(\mathrm{Y} \leq y) = P_Y(X_1 \cup X_2) \tag{60}$$

where $P_Y(X_1 \cup X_2)$ is the event, conditioned by $X_1 + X_2 \leq y$. If the *pdfs* of the X_1, X_2, Y variables are used then the Eq.(60) becomes:

$$F_Y(y) = \int_{-\infty}^{Y} F_Y(t)dt = \int_{x_1}\int_{x_2} f_{X_1}(x_1)\cdot f_{X_2}(x_2)dx_1dx_2 \quad (61)$$

$$(x_1 + x_2 < y)$$

Substituting the variable *u for* x_1 *and* v for x_1+x_2 in Eq.(61)

$$F_Y(y) = \int_{-\infty}^{Y} f_Y(t)dt = \int_{-\infty}^{y}[\int_{-\infty}^{+\infty} f_{X_1}(u)\cdot f_{X_2}(v-u)du]dv \quad (62)$$

Accordingly,

$$f_Y(y) = \int_{-\infty}^{+\infty} f_{X_1}(u)\, f_{X_2}(y-u)du = f_{X_1} \otimes f_{X_2(y)} \quad (63)$$

where $f_Y(y)$ is the convolution of the of $f_{x_1}$and f_{x_2} *pdfs.*

The convolution of two functions is a mathematical operator which has specific properties as commutatively and associatively, but the most important property lies in the fact that the Fourier transform of $f_{X_1} \otimes f_{X_2}$ is the product of the Fourier transforms of the respective functions.

Based on Eq.(63) one may supposes that the *pdf* of the variable $Y_n(y)=X_1+X_2+....X_n$ is:

$$F_{Y_n}(y) = f_{X_1} \otimes f_{X_2} \dots\dots \otimes F_{X_n}(y) \quad (64)$$

where: $f_{X_i}, i = \overline{1,n}$ are the *pdfs* of the X_i variable.

The validity of Eq.(64) is proved by mathematical induction method. Thus, the above assumption is valid on condition that the *pdf* of the variable $Y_{n+1}(y)=X_1+X_2+....X_n+X_{n+1}$ is of the same form as that given by Eq.(64). To prove that, Y_{n+1} is written as:

$$Y_{n+1}(y)=X_1+X_2+....X_n+X_{n+1}=Y_n+Y_{n+1} \quad (65)$$

Accordingly,

$$f_{Y_{n+1}}(y) = f_{Y_n} \otimes f_{X_{n+1}}(y) = f_{X_1} \otimes f_{X_2} \dots .\otimes f_{X_n} \otimes f_{X_{n+1}} \quad (66)$$

Consequently, , Eq.(66) proves that Eq. (64) is valid for any *n*. In the case $f_{X_i}=f_X$, $i = \overline{1,n}$, the *pdf* of Y_n variable is the convolution product of *n* identical functions, denoted as :

$$f_{Y_n}(y) = f_X^{\otimes n}(y) \quad (67)$$

The *pdf* of a Y_n variable defined as: $Y_n= a_1X_1+a_2X_2+......+a_nX_n$ where X_i, $i = \overline{1,n}$ are random variables, a_i, $i = \overline{1,n}$ are real numbers, can be calculed in two steps. In the first step, the variables $Z_i =a_iX_i$ are introduced and then their *pdfs* are calculated as;

$$f_{Z_i}(Z_i) = \frac{1}{|a|} f_{X_i}\cdot(\frac{x_i}{a_i}) \quad (68)$$

Next, $f_{Y_n}(y)$ is calculated using the Eq. (64):

$$f_{Y_n}(y) = f_{Z_1} \otimes f_{Z_2} \dots .\otimes f_{Z_n}(y) \quad (69)$$

Note: The variables: $Y_n = \sum_{i=1}^{n} X_i$, with X_i=X, $i = \overline{1,n}$ and Y_n=$n \cdot X$ are diffrent.

The variable $\bar{\bar{X}}_n$, assigned to the mean of n numerical results obtained in repeatability conditions, also called as sample mean variable, is the typical variable to which the linear compound variables theory is applied. Thus, the $\bar{\bar{X}}_n$ has the expression:

$$\bar{\bar{X}}_n = \frac{1}{n}(X_1 + X_2 + \cdots X_n) = \frac{1}{n} \cdot Y_n \tag{70}$$

where X_i, $i = \overline{1,n}$, are the variable assigned to each measurement.

The pdf of $\bar{\bar{X}}$ is:

$$f_{\bar{\bar{X}}_n}(\bar{x}) = n \cdot f_{Y_n}(\bar{\bar{x}} \cdot n) = n \cdot f_X^{\otimes n}(n \cdot \bar{\bar{x}}) \tag{71}$$

If $f_X(x)$ is known then the experimental mean distribution can be calculated, and subsequently, the dispersion of the experimental mean around the conventional true mean μ of X can be estimated as well.

3.3. The probability density function of Y=X² variable

Consider a random variable X with $f_X : \mathbb{R} \to \mathbb{R}$ and a variable Y=X^2. By its definition Y has the following *cdf*:

$$F_Y(y) = \begin{cases} 0\,;\ y < 0 \\ F_Y(y) = \int_{u=0}^{y} f_Y(u)\, du\ ;y \geq 0 \end{cases} \tag{72}$$

where $f_Y(y)$ is the *pdf* of Y for $y \geq 0$.

The condition $u \leq y$ implies that $x^2 \leq y$ i.e $X \in [-\sqrt{y}, \sqrt{y}]$, respectively. Accordingly, the probability that $u \in [0,y]$ is equal to the probability that $X \in [-\sqrt{y}, \sqrt{y}]$ i.e.

$$P(Y \leq y) = P(-\sqrt{y} \leq x \leq \sqrt{y}) = F_x(\sqrt{y}) - F_x(-\sqrt{y}) \tag{73}$$

The Eq.(73) may be expressed as:

$$\int_{-\sqrt{y}}^{\sqrt{y}} f_X(x)dx = \int_{=\sqrt{y}}^{0} f_X(x)dx + \int_{0}^{\sqrt{y}} f_X(x)dx = I_1 + I_2 \tag{74}$$

where $I_1 = \int_{x_1=-\sqrt{y}}^{x_2=0} f_X(x)dx\,, I_2 = \int_0^{\sqrt{y}} fx(x)dx$

Substituting the variable $-\sqrt{t}$ for X into I_1

$$I_1 = \int_y^0 f_X(-\sqrt{t})\left(-\frac{1}{2\sqrt{t}}dt\right) = \frac{1}{2}\int_0^y \frac{f_X(-\sqrt{t})}{\sqrt{t}}dt \tag{75}$$

Likewise, if in I_2 is applied the substitutions $\sqrt{t}$, for x then

$$I_2 = \frac{1}{2}\int_0^y \frac{f_X(\sqrt{t})}{\sqrt{t}}dt \tag{76}$$

From Eqs. (74), (75) and (76) it is deduced that:

$$f_Y(y) = (1/(2\sqrt{y})) \cdot [fx(-\sqrt{y}) + fx(\sqrt{y})] \tag{77}$$

The repartition density of X^2 variable differs significantly from that of X. For example, if X is a variable with the *N(0,1) pdf* then the the *pdf* of variable $Y = X^2$ is:

$$f_Y(y) = \frac{1}{2\sqrt{y}} y^{-1/2} e^{-y/2} \tag{78}$$

which is of Weibull type.

3.4. The probability distribution function of $Y = \sqrt{X}$ variable

Consider X as a random continuous and positive variable with the *pdf* fx(x): $\mathbb{R} \to \mathbb{R}$. The $Y=\sqrt{X}$ variable has the value $y=\sqrt{x}$ when $X=x$. The *cdf* of Y is:

$$F_Y(y) = \int_0^y f_Y(u) = P_Y(Y \leq y) \tag{79}$$

On the other hand, $P_Y(Y \leq y)$ is equal to $P_X(X \leq x)$ i.e.

$$\int_0^y f_Y(u)du = \int_0^x f_X(t)dt \tag{80}$$

If on the right hand side of Eq.(80) one substitutes v^2 for t then

$$\int_0^y f_Y(u)du = \int_0^y 2v f_X(v^2)dv \tag{81}$$

Accordingly,

$$f_Y(y) = 2y f_X(y^2) \tag{82}$$

Generally, Eq.(79) is used for estimating the *pdf* of the standard deviation when the *pdf* of sample dispersion (variance) is known.

3.5. The probability distribution function of the ratio of two distributions

Let be the two random and independent variables X_1 and X_2 and their *pdfs* $f_{X_1}(x_1)$ and $f_{X_2}(X_2)$, respectively defined on $\mathbb{R}$. The variable $Y=X_1/X_2$ has the repartition function:

$$F_Y(y) = \int_{-\infty}^y f_Y(t)dt = P_Y(Y \leq y) \tag{83}$$

Substituting u for x_2 and $u \cdot v$ for x_1 in Eq.(83)

$$f_Y(y) = \int_{v=0}^y \int_{-\infty}^{+\infty} f_{X_1}(u \cdot v) \cdot f_{X_2}(u) \cdot \left| D\left(\frac{x,y}{u,v}\right) \right| dudv \tag{84}$$

where $D\left(\frac{x,y}{u,v}\right) = \begin{vmatrix} \frac{dx_1}{du} & \frac{dx_1}{dv} \\ \frac{dx_2}{du} & \frac{dx_2}{dv} \end{vmatrix} = \begin{vmatrix} v & u \\ 1 & 0 \end{vmatrix} = -u$ is the Jacobian of the coordinate transformation

From Eqs.(83) and (84)::

$$f_Y(y) = \int_{-\infty}^{+\infty} f_{X_1}(u \cdot v) \cdot f_{X_2}(u) \cdot \left| D\left(\frac{x,y}{u,v}\right) \right| dudv \tag{85}$$

The *pdf* of $Y=X_1/X_2$ variable with X_1 and X_2 of $N(0, 1)$ type is:

$$f_Y(y) = \int_{-\infty}^{+\infty} \frac{1}{\sqrt{2\pi}} \cdot e^{-\frac{u^2 y^2}{2}} \cdot \frac{1}{\sqrt{2\pi}} e^{-u^2} |u| du = \frac{1}{\pi(1+y^2)} \quad (86)$$

Eq.(86) shows that the *pdf* of the ratio of two variables with standard normal distribution is the Cauchy standard distribution.

3.6. General approach for deriving the *pdf* of the sample mean variable

As reported in literature [2, 8] the sample mean of a sample size "*n*" i.e. $\{x_1, x_2, ..., x_n\}$ is an estimator for the population mean. On the other hand, the mean has two different meanings: 1) numeric value of the sample mean calculated from observed values of the sample and 2) a function of random variables from a random sample. This subchapter addresses the *pdf* assigned to the mean of the outcomes of n repeated tests. Therefore, the mean as a function is a sum of *n* identical functions divided by *n*. The *pdf* of X, as it was derived in § 3.1, is:

$$f_{\bar{X}}(\bar{x}) = n f^{\otimes n}(n\bar{x}) \quad (87)$$

In the next three paragraphs the $f_{\bar{X}}$ will be deducted for Gaussian, uniform and Cauchy *pdfs*.

3.7. The probability distribution function of mean of "*n*" Gaussian variables

Let Y_n be a compound variable of *n* identical Gaussian variable *X* defined as $Y_n = X_1+X_2+...+X_n$, where: $X_i \equiv X, i = \overline{1,n}$. The *pdf* of *X* is:

$$f_{(x;\mu,\sigma^2)} \equiv f(x) = \frac{1}{\sigma\sqrt{2\pi}} = e^{-\frac{(x-\mu)^2}{2\sigma^2}} \quad (88)$$

As proven in § 3.1.2., the *pdf* of $Y_2 = X_1 + X_2$ is:

$$f_{Y_2}(y) = \int_{-\infty}^{+\infty} f_{X_1}(y-u) \cdot f_{X_2}(u) du \quad (89)$$

Eq.(89) shows that the *pdf* of Y_2 is of Gaussian type with the mean $\mu_2 = 2\mu$ and standard deviation $\sigma_2 = \sqrt{2} \cdot \sigma$. Based on the above result, let's assume that the Y_n variable has a *pdf* of the form

$$f_{Y_n}(y) = \frac{1}{\sqrt{n}\,\sigma \cdot \sqrt{2\pi}} \cdot e^{-\frac{(\sigma-n\mu)^2}{2n\sigma^2}} \quad (90)$$

According to the complete induction method the above assumption is true if on the basis of Eq.(90) it can be proven that the *pdf* of the Y_{n+1} variable is:

$$f_{Y_{n+1}}(Y) = \frac{1}{\sigma\sqrt{n+1} \cdot \sqrt{2\pi}} \cdot e^{-\frac{(y-(n+1)\mu)^2}{2(n+1)\sigma^2}} \quad (91)$$

The variable $Y_{n+1} = Y_n + X_{n+1}$, therefore *pdf* of Y_{n+1} may be written as;

$$\frac{1}{2\pi\sqrt{n}\sigma^2} \int_{-\infty}^{+\infty} e^{-\frac{(y-u-\mu)^2}{2\sigma^2}} \cdot e^{\frac{-(u-n\mu)}{2n\sigma^2}} du = \frac{e^{-\frac{(y-(n+1)\mu)^2}{2(n+1)\sigma^2}}}{\sigma \cdot \sqrt{n+1} \cdot \sqrt{2\pi}} \quad (92)$$

Eq.(92) proves that Eq.(90) is true and permits to state that the *pdf* of a sum of "n" identical Gaussian variable is of Gaussian type having the mean $\mu_n = n \cdot \mu$ and the standard deviation $\sigma_n = \sqrt{n} \cdot \sigma$

As it was proven in § 3.1. the *pdf* of the sample mean variable $\bar{X}$ is:

$$f_{\bar{x}}(\bar{x}) = n \cdot f_{Y_n}(n, \bar{x}) = n \cdot \frac{1}{\sigma\sqrt{n} \cdot \sqrt{2\pi}} \cdot e^{\frac{-(n\bar{X}-n\mu)}{2n\sigma^2}} = \frac{e^{-\frac{(\bar{x}-\mu)^2}{2\sigma_{\bar{x}}^2}}}{\sigma_{\bar{x}} \cdot \sqrt{2\pi}} \tag{93}$$

where $\sigma_{\bar{x}} = \sigma/\sqrt{n}$ is the standard deviation of the mean when the sample size is n.

Eq.(93) shows that the mean values ($\bar{x}$) are centered on the population mean μ and their standard deviation is $\sqrt{n}$ times smaller than standard deviation of the population.

The mean and the variance of X could be easily estimated on the base of mean operator (M) and of variance V one applied to a vector of statistical variable, i.e.[6]

$$M(\sum_{i=1}^{n} a_i \cdot X_i) = a_i \sum_{i=1}^{n} M(X_i) = \frac{1}{n} \cdot n \cdot \mu = \mu \tag{94}$$

$$V(\sum_{i=1}^{n} a_i \cdot X_i) = a_i^2 \sum_{i=1}^{n} V(X_i) = \frac{1}{n^2} n\sigma^2 = \frac{\sigma^2}{n} \tag{95}$$

where: $a_1 = 1/n, i = \overline{1,n}$, but these operations has no meaning for the experimentalist.

3.8. The probability distribution function of mean of some uniform variable

In real world testing situations are often found where there is no knowledge about the *pdf* of the measurand. In such cases, the experimentalists have to consider that the *pdf* assigned to the measurand is of the uniform type. The same, metallurgical practice implies statistical modeling using additive uniform variable. Simple examples are: weight or length of a chain with n links, strength resistance of a series of n bars, fiability assessment of a product composed of n parts. According to §2.3.2 any *updf* may be related to a so called standard *updf* having the width 2, μ=0 and SD=$\sqrt{3}/3$

Based on the above consideration this section addresses only standard *updf* given by Eq.(34).

As was derived in §3. 2.1, the *pdf* assigned to arithmetic mean of n outcomes is:

$$f_{\bar{\bar{X}}} = n \cdot f_X^{\otimes n}(x) \tag{96}$$

where $X_i = X, i = \overline{1,n}$ .The first step in deriving the *pdf* of X is to calculate the $f_{(X)}^{\otimes n}$. The expression of $f_{(x)}^{\otimes n}$ will be derived by recurrence approach i.e $f^{\otimes((k+1)}=f^{\otimes k} \otimes f(x)$. Thus, the two fold convolution of f is:

$$f_{(x)}^{\otimes 2} = \int_{-\infty}^{+\infty} f_{X_1}(x-u) \cdot f_{X_2}(u)du = \int_{=\infty}^{+\infty} f_{X_1}(u) \cdot f_{X_2}(u-x)du \tag{97}$$

where $f_X(u-x)$ is equivalent with a translation of the graph of $f_{X_2}$with x related to the origin of coordinate. The value of the above integral is proportional to the overlapping area of the graphs of the X_1 and X_2 *pdfs*, which is:

$$f^{\otimes 2}(x) = \begin{cases} \frac{1}{2^2}(2-|x|); x \in [-2,2] \\ 0\,; \notin [-2,2] \end{cases} \quad (98)$$

The $f^{\otimes 2}$ is known and as Simpson distribution [9]. The convolution of two uniform variables gives a triangle distribution with the same height as convoluted *pdf* but two times larger.

The $f^{\otimes 3}_{(x)}$ is calculated as:

$$f^{\otimes 3}_{(x)} = \int_{-\infty}^{+\infty} f^{\otimes 2}(u) \cdot f(u-x)dx = \begin{cases} \frac{1}{2^3}(3-x^2)\,; x \in [-1,1] \\ \frac{1}{2^4}(3-|x|)^2;\ x \in [-3-1) \cup (1,3]\ 1 \le |x| \le 3 \\ 0\,;\ |x| > 3 \end{cases} \quad (99)$$

$f^{\otimes 3}$ is very important because it is the keystone from where the *pdf* of a sum of identical uniform distribution turns in a polynomial shape (Figure 4). The first order derivative of $f^{\otimes 3}$ is continuous in $X = \pm 1$ but the second order one is not.

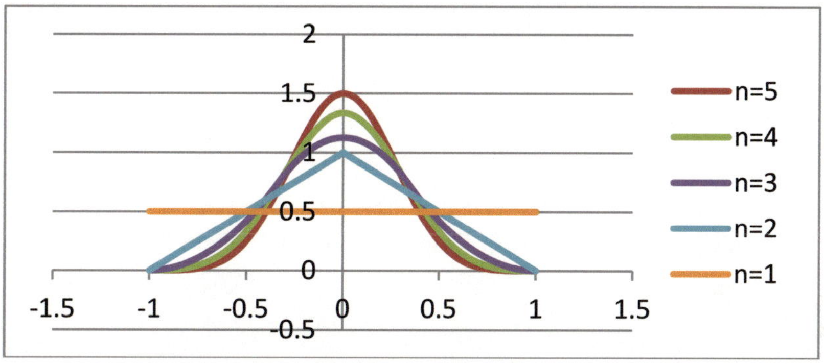

Figure 4. The graphs of the sample means *pdfs* for $n=\overline{1,5}$

Thus, at $x_{1,2} = \pm 1$ are two inflection points. In the same way as for $f^{\otimes 3}_{(x)}$,$f^{\otimes 4}$ can be calculate as;

$$f^{\otimes 4}_{(x)} = \begin{cases} \frac{1}{2^4 \cdot 3!}(32 - 12x^2 + 3|x|^3);\ |x| < 2 \\ \frac{1}{2^4 \cdot 3!}(4-|x|)^3; 2 \le |x| \le 4 \\ 0; x > 4 \end{cases} \quad (100)$$

The $f^{\otimes 5}$ *pdf* is deduced as:

$$f^{\otimes 5}_{(X)} = \begin{cases} \frac{2}{2^5 \cdot 4!}(115 - 30x^2 + 3X^4)\,;\ |x| < 1 \\ \frac{4}{2^5 \cdot 4!}(55 - 10|x| - 30X^2 + 10|x|^3 - X^4)\,; 1 \le |x| < 3 \\ \frac{4}{2^5 \cdot 4!}(5-|x|)^4\,;\ 3 \le |x| \le 5 \\ 0\,;\ |x| > 5 \end{cases} \quad (101)$$

The expressions of the k-convolved standard *updfs* are difficult to be calculated for k>5, but one can get help using the general form of a sum of n identical standard uniform variables given by Renyi[9]:

$$f_{(x)}^{\otimes n}\begin{cases}\frac{1}{(n-1)!2^n}\sum_{i=0}^{\hat{n}(n,x)}(-1)^1\cdot C_n^i(x+n-2i)^{n-1};\ -n\le x\le n\\ 0;\qquad otherwise\end{cases} \tag{102}$$

where $\hat{n}(n,x)=\left[\frac{x+n}{2}\right]$ is the largest integer less than $\frac{x+n}{2}$

The *pdf* of the sample mean of the 5 uniform distributed outcomes obtained in repetitive or reproductive condition, denoted $f_{M5}(m)=5f_{(5m),}^{\otimes 5}$ is

$$f_{M5}(m)=\begin{cases}\frac{10}{2^5\cdot 4!}(115-6\cdot 5^3m^2+3\cdot 5^4\cdot m^4);\ |m|<\frac{1}{5}\\ \frac{20}{2^5\cdot 4!}(55-50|m|-6\cdot 5^3m^2+2\cdot 5^4\cdot|m|^3-5^4\cdot m^4);\frac{1}{5}\le|m|<\frac{3}{5}\\ \frac{5^5}{2^5\cdot 4!}(1-|m|)^4;\ \frac{3}{5}\le|m|\le 1\\ 0;\ |m|>1\end{cases} \tag{103}$$

The graphs of the sample means for $n=\overline{2,5}$ are given in Figure 4. A special attention is drawn to $f_{M5}(m)$ because it is appropriate for the hardness test where the standard recommends five reproductive measurements. Based on the f_{M5} there were calculated: $P\left(|x|<\frac{1}{5}\right)\cong$ 25,5%; $P\left(\frac{1}{5}\le|x|<\frac{3}{5}\right)=44.8\%$; $P\left(\frac{3}{5}\le|x|<1\right)=0.7\%$. On the other side, the probability that the mean lies in the interval $(-\frac{1}{10},\frac{1}{10})$ is about 29,5% while in $\left(-\frac{2}{10},-\frac{1}{10}\right]\cup\left[+\frac{1}{10};\frac{2}{10}\right)$ is 25,5 %. Thus, the probability that the mean depart from zero decreases relative slowly, not so rapidly as is argued elsewhere [10]. Based on the convolution approach or on Renyi's formula the experimentalist could calculate the *pdf* of the sample mean for its own n number of repeated tests.

3.9. The *pdf* of mean of some Cauchy distributed variable

Let us consider two Cauchy variable X_1 and X_2 whose *pdfs* defined on ℝ are:

$$f_{X_1}=\frac{a_1}{\pi(a_1^2+x^2)}\text{ and }f_{X_2}(x)=\frac{a_2}{\pi(a_2+x^2)} \tag{104}$$

The *pdf* of the variable $Y_2=X_1+X_2$ is:

$$f_{Y_2}(X)=\int_{-\infty}^{+\infty}fx_1(x-u)\,fx_2(u)=\frac{a_1+a_2}{\pi[(a_1+a_2)^2+x^2]} \tag{105}$$

In testing practice X_1 and X_2 are the same i.e. $a_1=a_2=a$ therefore f_{Y_2} is:

$$f_{Y_2}(x)=\frac{2a}{\pi[(2a)^2+x^2} \tag{106}$$

There expression of n-fold convolved Cauchy *pdfs* is :

$$f_{Y_n}(\mathrm{x})=\frac{na}{\pi[(na)^2+x^2} \tag{107}$$

The *pdf* assigned to the mean of a set of *n* outcomes having Cauchy distribution is:

$$f_{M_n}(m) = n \cdot f_{Y_n}(nm) = \frac{n^2 a}{\pi(na)^2+(nm)^2} = \frac{a}{\pi(a^2+m^2)} \tag{108}$$

Eq.(108) shows that the *pdf* of mean is identical with that of the measurand. As a paradox, repeating a test many times on a measurand whose *pdf* is of Cauchy type is of no use. Since the dispersion of the Cauchy distribution is infinity then it should be avoided being assigned to a measurand.

3.10. Student's or t-distribution

The Student's distribution, also referred to as t-distribution, is used on a large scale to test the exactness of a set of numerical outcomes of repeated tests when μ and σ are not known. The power of t-tests consists in making use of μ and σ as hidden or implicit variable for statistical inference while they remain unknown. Thus, let be a set of n outcomes $\{x_1, x_2,...,x_n\}$ whose mean $\bar{x}$ and sample dispersion s^2 are, respectively

$$\bar{x} = \frac{1}{n}\sum_{i=1}^{n} x_i \tag{109}$$

$$s^2 = \frac{1}{n-1}\sum_{i=1}^{n}(x_i - \bar{x})^2 \tag{110}$$

The experimentalist is concerned about the exactness of $\bar{x}$, i.e. the accuracy $\bar{x} - \mu$ related to the standard deviation of mean $s_{\bar{x}} = s/\sqrt{n}$. Thus, the parameter $t = (\bar{x} - \mu)/s_{\bar{x}}$ was found to be the best estimator for the case of a test with unknown μ and σ. Fortunately, a t-parameter can be written as:

$$t = \left(\frac{\bar{x}-\mu}{\frac{\sigma}{\sqrt{n}}}\right) / \left(\frac{\frac{S_{\bar{x}}}{\sqrt{n}}}{\frac{\sigma}{\sqrt{n}}}\right) = \left(\frac{\bar{x}-\mu}{\sigma_{\bar{x}}}\right) / \left(\frac{S_{\bar{x}}}{\sigma_{\bar{x}}}\right) \tag{111}$$

The *pdf* of the variable *T* assigned to *t* values will be derived as the ratio of the variables *Z* assigned to $(\bar{x} - \mu)/\sigma_{\bar{x}}$ and R to $(s_{\bar{x}}/\sigma_{\bar{x}})$. The Z variable has a *pdf* of *N(0,1)* type while the *pdf* of *R* will be defined a little bit later. Before proceeding to derive the expression of the *pdf* of *R* variable, let make some reconsiderations about the actual way of deriving the *pdf* of S^2. Thus, on the basis of the well known Eq.(110) the variable S^2 assigned to s^2 is considered as:

$$S^2 = \frac{(n-1)}{\sigma^2} {Y_n}^2 \tag{112}$$

where $Y_n^2 = (X_1 - \mu)^2 + \cdots + (X_n - \mu)^2$

Here it is considered that the above issue is quite unproductive because the *(n-1)* factor replaces the *n* factor in Eq.(107) just because s_n^2 approximates better the sample variance related to μ. i.e.

$$\frac{1}{n-1}\sum(x_i - \bar{x})^2 \cong \frac{1}{n}\sum_{i=1}^{n}(x_i - \mu)^2 = \frac{\sigma^2}{n}\sum_{i=1}^{n}\left(\frac{x_i-\mu}{\sigma}\right)^2 \tag{113}$$

Therefore the variable attached to S^2 should be:

$$S^2 = \frac{\sigma^2}{n}\sum_{i=1}^{n}\phi_i(n_i) = \frac{\sigma^2}{n}\phi_n \tag{114}$$

where $\phi_i = N(0,1)$, $i = \overline{1,n}$ and $\phi_n = \sum_{i=1}^{n}\phi_i$.

The *pdf* of ϕ_n, as it was shown in § 3.3.), is:

$$f_{\phi_n}(u) = \frac{u^{\frac{n}{2}-1}e^{-\frac{n}{2}}}{2^{\frac{n}{2}}\Gamma(\frac{n}{2})} \tag{115}$$

The *pdf* of S_n^2 variable is:

$$f_{S_n{}^2}(w) = \frac{n}{\sigma^2}f_{\phi_n}\left(\frac{nw}{\sigma^2}\right) = \frac{n}{\sigma^2}\frac{\left(\frac{nw}{\sigma^2}\right)^{\frac{n}{2}-1}e^{-\frac{nw}{2\sigma^2}}}{2^{\frac{n}{2}}\Gamma(\frac{n}{2})} \tag{116}$$

The pdf of $S_n = \sqrt{S_n{}^2}$ variable is:

$$f_{S_n}(s) = 2s\cdot f_{S_n{}^2}(s^2) = \frac{2s\left(\frac{n}{\sigma^2}\right)\left(\frac{nS^2}{\sigma^2}\right)^{\frac{n}{2}-1}e^{-\frac{nS^2}{2\sigma^2}}}{2^{\frac{n}{2}}\Gamma(\frac{n}{2})} \tag{117}$$

The *pdf* of the variable $R = S_n/\sigma$ is:

$$f_R(r) = \sigma f_{S_n}(r\sigma) = \frac{2nr(n)^{\frac{n}{2}-1}r^{n-2}e^{-\frac{nr^2}{2}}}{2^{\frac{n}{2}}\Gamma(\frac{n}{2})} = \frac{2n^{\frac{n}{2}}r^{n-1}e^{-\frac{nr^2}{2}}}{2^{\frac{n}{2}}\Gamma(\frac{n}{2})} \tag{118}$$

The *pdf* of $T = \emptyset_n/R$, according to §3.1, can be calculated as:

$$f_T(t) = \int_0^{\infty}\frac{e^{-\frac{(tu)^2}{2}}}{\sqrt{2\pi}}\frac{2n^{\frac{n}{2}}u^{n-1}e^{-\frac{nu^2}{2}}}{2^{\frac{n}{2}}\Gamma(\frac{n}{2})}u\cdot du = \frac{n^{\frac{u}{2}}}{\sqrt{2\pi}2^{\frac{n}{2}}\Gamma(\frac{n}{2})}\int_0^{\infty}e^{-\frac{u^2(t^2+n)}{2}}u^{n-1}2u\cdot du \tag{119}$$

Substituting x for $[u^2(t^2+n)]/2$ in Eq.(119)

$$f_T(t) = \frac{n^{\frac{n}{2}}}{\sqrt{2\pi}2^{\frac{n}{2}}\Gamma(\frac{n}{2})}\int\frac{x^{\frac{n-1}{2}}2^{\frac{n-1}{2}}e^{-x}}{(t^2+n)^{\frac{n-1}{2}}}\frac{1}{(t^2+n)}\cdot dx = \frac{\Gamma\left(\frac{n+1}{2}\right)}{\sqrt{\pi n}\Gamma(\frac{n}{2})}\left(1+\frac{t^2}{n}\right)^{\frac{-(n+1)}{2}} \tag{120}$$

The expression of $f_T(t)$ is consistent with those given in different works [6, 8] if n is replaced by υ which is the so called number of degrees of freedom (*ndf*) of the variable under consideration.

The derivation of $f_T(t)$, frequently denoted as *f(t)*, is important for two reasons: a) the Student's *pdf* must be applied only for a set of identical Gaussian variables or n repeated test when the *pdf* of the measurand is of normal (Gaussian) type; b) to correctly establish the *ndf*.

The *cdf* of *T*, $F_T(t)$, cannot be estimated as analytical primitive of $f_T(t)$, but its values were tabulated and can easily be found in the open literature [5, 6, 8].The arguing that $\upsilon=n$ is very important when n is 2, 3, 4 because the $t_n(0; 0.05)$ varies significantly in this range. Thus, it is a matter of evidence that many times 3 repeated measurements are considered to be enough.

In this case, using $t_2(0.005)=4.303$ instead of $t_3=(0.05)=3.182$ increases the expanded uncertainty significantly.

3.11. Fisher-Snedecor distribution

The proficiency testing (PT) is a well defined procedure used for estimating the performances of the collaborative laboratories [11, 12]. ANOVA (ANalysis Of VAriance) is one of the methods used for analysis of sample variances reported by the laboratories. ANOVA is based on Fisher-Snedecor distribution of the two sample variances of a measurand X. Thus, if for the same sample a laboratory labeled A, gives a sample dispersion.

$$s_A{}^2 = \frac{1}{n_1-1}\sum_{i=1}^{n_1}(x_i - \overline{x_1})^2 \tag{121}$$

while B laboratory gives:

$$s_B{}^2 = \frac{1}{n_2-1}\sum_{j=1}^{n_2}\left(x_j - \overline{x_2}\right)^2 \tag{122}$$

where x_i, $i = \overline{1, n_1}$ and $x_j = \overline{1, n_2}$ are the outcomes obtained for the same X measurand by the laboratories A and B, respectively.

As it was argued in §3.4, to the $s_A{}^2$ and $s_B{}^2$ may be assigned the variables $S_A{}^2$ and $S_B{}^2$ whose *pdfs* are, respectively:

$$f_{S_A{}^2}(u) = \frac{n_1^{\frac{n_1}{2}} \cdot u^{\left(\frac{n_1}{2}-1\right)} \cdot e^{-\frac{n_1 u}{2\sigma^2}}}{\sigma n_1 \cdot 2^{\frac{n_1}{2}} \cdot \Gamma\left(\frac{n_1}{2}\right)} \tag{123}$$

$$f_{S_B{}^2}(v) = \frac{n_2^{\frac{n_2}{2}} \cdot v^{\left(\frac{n_2}{2}-1\right)} \cdot e^{-\frac{n_2 v}{2\sigma^2}}}{\sigma n_2 \cdot 2^{\frac{n_2}{2}} \cdot \Gamma\left(\frac{n_2}{2}\right)} \tag{124}$$

The *pdf* of the variable ratio $S^2{}_A/S^2{}_B=F$ is:

$$f_F(x) = \int_0^\infty f_{S_A{}^2}(xy) \cdot f_{S_B{}^2}(y) \cdot y \cdot dy = \frac{n_1^{\frac{n_1}{2}} \cdot n_2^{\frac{n_2}{2}} \cdot \int_0^\infty (xy)^{\frac{n_1}{2}-1} \cdot y e^{-\frac{y}{2}\left(\frac{n_1 X + n_2}{\sigma^2}\right)} dy}{\sigma^{n_1} \cdot 2^{\frac{n_1}{2}} \cdot \Gamma\left(\frac{n_1}{2}\right) \cdot \sigma^{n_2} \cdot 2^{\frac{n_1}{2}} \cdot \Gamma\left(\frac{n_1}{2}\right)} =$$
$$\frac{n_1^{\frac{n_1}{2}} \cdot n_2^{\frac{n_2}{2}} \cdot x^{\left(\frac{n_1}{2}-1\right)}}{\sigma^{n_1+n_2} \cdot 2^{n_1+n_2} \cdot \Gamma\left(\frac{n_1}{2}\right) \cdot \Gamma\left(\frac{n_2}{2}\right)} \cdot \int_0^\infty y^{\frac{n_1+n_2}{2}-1} \cdot e^{-y\left(\frac{n_1 x + n_2}{2\sigma^2}\right)} dy \tag{125}$$

If the variable y is replaced by $u = y(n_1 x + n_2)/2\sigma^2$ in Eq.(125) then:

$$f_F(x) = \frac{n_1^{\frac{n_1}{2}} \cdot n_2^{\frac{n_2}{2}} \cdot x^{\left(\frac{n_1}{2}-1\right)}}{\sigma^{n_1+n_2} \cdot 2^{n_1+n_2} \cdot \Gamma\left(\frac{n_1}{2}\right) \cdot \Gamma\left(\frac{n_2}{2}\right)} \cdot \frac{2^{\frac{n_1+n_2}{2}} \sigma^{(n_1+n_2)}}{(n_1 X + n_2)^{\frac{n_1+n_2}{2}}} \cdot \int_0^\infty u^{\frac{n_1+n_2}{2}-1} e^{-u} du = \frac{\left(\frac{n_2}{n_1}\right)^{\frac{n_1}{2}} \cdot \Gamma\left(\frac{n_1+n_2}{2}\right)}{\Gamma\left(\frac{n_1}{2}\right) \cdot \Gamma\left(\frac{n_2}{2}\right)} \cdot \frac{x^{\frac{n_1}{2}-1}}{\left(x+\frac{n_2}{n_1}\right)^{\frac{n_1+n_2}{2}}} =$$
$$\left(\frac{n_2}{n_1}\right)^{\frac{n_1}{2}} \cdot B\left(\frac{n_1}{2}, \frac{n_2}{2}\right) \cdot x^{\frac{n_1}{2}-1} \cdot \left(x + \frac{n_2}{n_1}\right)^{-\frac{n_1+n_2}{2}} \tag{126}$$

The derived expression for $f_F(x)$ is identical to that given by Cuculescu, I., [5]. Herein, again arises the problem where *ndf* is *n*-1 or *n*. By my opinion *ν*=*n*. The *cdf* of Fisher-Snedecor distribution is not an elemental analytical form, but its value for different significance level (*sl*) and different n_1, n_2, $F(sl,n_1,n_2)$, are tabulated and can be easily found in open literature [5, 6, 8]. The way of using Fisher-Snedecor distribution as *F*-test consists of comparing the obtained value of $s^2_A/s^2_B=f_e$ with the value of $F(sl; n_1,n_2)$ taken from a Fisher *cdf* table. If $f_e \leq F(sl; n_1, n_2)$, then the s^2_A and s^2_B are consistent, otherwise one of them is a straggler or an outlier.

Note. A straggler should be considered a correct item, but a statistical outlier is discharged [11].

4. Measurement uncertainty estimation

4.1. General concepts regarding measurement uncertainty estimation

Apparently the terms "test result" and "measurement result" have the same meaning i.e. a numeric outcome of a measurement process. But in metrology a measurement is defined as a "set of operations having the object of determining a value of a quantity" [13, 14] while a test is defined as a "technical operation that consists of the determination of one or more characteristics of a given product, process or service according to a specific procedure" [14]. Thus, a test is a measurement process well documented, fully implemented and permanently supervised. In a test process, the environmental and operational conditions will either be mentioned at standard values or be measured in order to apply correction factors and to express the result in standardized conditions. Besides the rigorous control of the "test conditions", a test result bears an intrinsic inaccuracy depending on the nature of the measurand, on the performance of the method and on the performances of the equipment. Thus, the entire philosophy of the metrology is based on the fact that the true value of the measurand remains unknown to a certain extent when it is estimated based on a set of test outcomes. In this sense, the measurement uncertainty (*MU*) is defined as an interval about the test result, which may be expected to encompass a large fraction of values that could reasonably be attributed to the measurand. The half width of this interval is called expanded uncertainty, denoted by *U* or *U(xx)*, where *xx* is the index of the level of confidence associated to the interval. The confidence level is frequently 95% but may be about 99% or 68% depending on the client requirements or on the test performances. According to EN ISO/IEC 17025 a testing laboratory should apply a procedure to estimate the *MU* [1]. In a similar manner, when estimating the *MU* all relevant uncertainty components in a given situation should be taken into account using the appropriate method of analysis. The sources of uncertainty include but are not limited to the reference standards and reference materials used, the method and equipment used, the environmental conditions, the properties and conditions of the item being tested and the operator. The degree of rigor of the uncertainty estimation depends on many factors as those above, but the test result quality is better as the *MU* is smaller. Thus, the *MU* is the most appropriate quantitative parameter for test result quality evaluation. As a consequence, the approaches of *MU* estimation were the subject to different committees of the renowned organizations such as ISO, ILAC, EA, EURACHEM, AIEA etc. Among the *MU* estimation approaches, the one of ISO given in GUM [2] may be considered as the master one. Accordingly, consistency with GUM is generally required for any particular procedure for *MU* estimation. The basic concepts of GUM are:

1. The knowledge about any influence factor of the measurand is in principle incomplete, therefore its action shall be considered stochastic following a *pdf*;
2. The expected value of the *pdf* is taken as the best estimate of the influence factor;
3. The standard deviation of the *pdf* is taken as the best estimate of the standard uncertainty associated to that estimate;
4. The type and the parameter(s) of the *pdf* have been obtained based on prior knowledge about the influence factors or by repeated trials of the test process.

The *MU* estimation procedure consists of two main steps. The first step, called formulation, consists of measurand description (physical and mathematical modeling), statistical modeling, input-output modeling and, finally, assigning a *pdf* to the measurand. The second step, called calculation, consists of deriving the *pdfs* for the test result estimation (mean, standard derivation, etc.) and the formulas or the algorithm for estimating the *MU* attributed to the test result.

4.2. Uncertainty estimation according to GUM

4.2.1. Formulation step

The formulation begins with the measurand definition. The *X* measurand may be classified as directly accessible to the measure or indirectly measured. The directly measurable measurands as length, mass, temperature etc. are not addressed here but those indirectly measured as elemental concentrations, Young modulus, hardness etc. Anyhow, a SI unit of measure shall be assigned to each measurand whenever it is possible, and also a reference for traceability. The indirect measurand may have a deterministic mathematical model as for Young modulus

$$E = \frac{F \cdot l_0}{S \cdot Dl} \tag{127}$$

where E is the Young modulus, F is the applied force (N), S is the area (m^2), l_o is the initial length of the specimen, Δl is the elongation (m).

Generally, the deterministic model of a measurand Y which is estimated based on the values of a set of measurands $X_i, i = \overline{1,n}$ is a function that described the relationship between outcome size and sizes of the inputs expressed as:

$$Y = f(X_1, X_2, \dots, X_n) \tag{128}$$

Eq.(125) is called the input-output model of the test process. A particular value y of Y is calculated based on the x_i values of $X_i: i = \overline{1,n}$ as:

$$y = f(x_1, x_2 \dots x_n) \tag{129}$$

But, from the statistical point of view the mathematical (statistical) model of the measurand Y is [11, 13]:

$$y = \mu + b + e \tag{130}$$

where μ is the general mean or the true value of Y, b is the systematic error or bias and e is the random error occurring in every measurement.

The bias b may be detected and corrected by statistical means. The random error e is caused by the whole set of the influence factors of the input measurands that form the so called uncertainty budget. The $X_i, i = \overline{1,n}$ measurands may be directly accessible to the measurement or may not but each X_i has its own uncertainty budget. Thus, the statistical model of the X_i ; $i=1,n$ are of the form:

$$X_i = \mu_i + b_i + e_i \tag{131}$$

where μ_i is the test value of X_i, b_i is the bias of X_i and e_i is the random error of X_i

Thus, the systematic errors and the random ones of the input measurands $X_i : i = \overline{1,n}$ are incorporated in the overall uncertainty of the output measurand Y. Each $X_i : i = \overline{1,n}$ has an uncertainty budget (UB) containing n_i influence factors F_{ij}, $i = \overline{1,n}$, denoted as $UB_{Xi} = \{ F_{ij} \}$; $j = \overline{1, n_i}$. In this instance, the uncertainty budget of Y, denoted as UB_Y, could be expressed as:

$$UB_Y = \cup_{i=1}^{n} (UB_{X_i}) \tag{132}$$

Eq.(132) shows that the design of an accurate uncertainty budget for a measurand needs advanced knowledge and extended data about the input measurands that, many times, are quite impossible to be achieved. In this context, the *pdf* of Y remains the most appropriate target for the experimentalist. The *pdf* assigned to the Y may be of Gaussian type if the influence factors are all of the Gaussian type or if its uncertainty budget contains more than 30 uncorrelated influence factors. Otherwise, to the Y shall be assigned a *pdf* having a less clustering tendency than the normal one. Same *pdfs* having less clustering tendency could be ordered as uniform, trapezoidal, Cauchy, etc. The assigning of an appropriate *pdf* to a measurand is a difficult task which has to be solved by the experimentalist. One of the arguments that underpins the previous affirmation consists the evidence that each measurand has its own variability which is enlarged by the testing process.

4.2.2. The calculation step

Suppose that the *pdf* of a measurand X, $f_X(x)$, is established. As it was shown in §3.1, the arithmetic mean of a set of test results carried on in repetitive or reproductive conditions is the best estimator for the conventional true value of the measurand. The variable assigned to the mean, M, is described by a linear model as:

$$M = (X_1 + X_2 + \ldots\ldots X_n) / n \tag{133}$$

where x_i, $i = \overline{1,n}$, are the statistical variable assigned to each repeated measurement.

The mean is a statistic [2, 5, 6,] with a *pdf* obtainable by a convolved product as it is shown in §3.2

$$f_M(y) = nf^{\otimes n}(ny) \tag{134}$$

Once having the $f_M(y)$ then it is quite easily to calculate the probability $p(|\bar{x} - \mu|) \leq \varepsilon$ i.e. to estimate the level of confidence for $\bar{x} - \varepsilon \leq \mu \leq \bar{x} + \varepsilon$.

On the other hand, the design of the mathematical model of the measurand and, subsequently, of the sample mean needs substantial scientific efforts and costs that may be prohibitive. As a consequence, many times laboratories adopt an alternative approach based on prior empirically achieved information.

4.3. The empirical estimation of *MU*

If a testing laboratory does not have a mathematical model as a basis for the evaluation of *MU* of the test results then it has to implement an empirical procedure for *MU* estimation. The flowchart of such a procedure is a stepped one [3]. Thus, the first step consists of listing those quantities and parameters that are expected to have a significant influence on the test result. Subsequently, the contribution of each influence factor to the overall *MU* is assessed. Based on the level of contribution to the overall *MU*, each factor shall be classified as significant or irrelevant. The irrelevant influence factors are discarded from the list. The equipment, the CRMs, the operator are among the most frequently considered as significant influence factors. If there is a lack of knowledge or prior data about an influence factor then it is strongly recommended to the laboratory to perform specific measurements to evaluate the contribution of that factor. If the contribution of an influence factor to the overall *MU* is estimated based on a set of outcomes whose variability is assigned exclusively to that factor then its contribution to the overall *MU* is considered of the A type. Else its contribution is of type B.

4.4. Combined uncertainty calculation

The combined uncertainty incorporates the contributions of the influence factors F_i, $i = \overline{1,n}$, to the overall *MU* i.e. uncertainties of the A type and of B type. The calculus of the combined uncertainty is based on error propagation law [2, 4, 8]. Metallurgical testing practice shows that the influence factors are about always considered as independent. Thus, the combined uncertainty is calculated as:

$$u_c^2 = u_A^2 + u_B^2 \tag{135}$$

where u_A and u_B are the combined uncertainty of A type and of B type, respectively.

The combined uncertainty of the type A is calculated as:

$$u_A = \left(\sum_{i=1}^{n} u_{A_i}^2\right)^{1/2} \tag{136}$$

where u_{Ai} is the uncertainty of type A assigned to the influence factor F_{ij}, $i = \overline{1,n}$.

The combined uncertainty of the type B is calculated as:

$$u_B = \left(\sum_{j=1}^{m} u_{B_i}^2\right)^{1/2} \tag{137}$$

where u_{Bi} is the uncertainty of type B assigned to the influence factor F_j, $j = \overline{1,m}$.

4.5. Estimation of the expanded uncertainty

The degree of confidence associated with the combined uncertainty (u_c) is considered, many times, as incompetent, therefore the uncertainty attributed to the measurand is increased to a certain extent in order to comply with a previously stated confidence level, usually 95%. The increased uncertainty is called expanded uncertainty and is usually obtained by multiplying u_c with a factor k, called coverage factor. For example, the k factor for a measurand with Gaussian *pdf* is 2 for a 95% confidence level, i.e. the expanded uncertainty for the 95% confidence level is $U=2u_c$. The value of k for a specific confidence level depends on the *pdf* type of the measurand and on the number of test results that were used to calculate the sample mean and sample standard derivation. If the *pdf* of the measurand is Gaussian or the number of test results exceeds 30 then it is justified to consider that the mean is normally distributed. If the *pdf* of the measurand is Gaussian but $n < 30$ then the Students' t-distribution shall be used.

According to [10,14], in many cases the uniform distribution may be assigned to the measurand. In this case, *U(95)* is calculated as:

$$U(95) = 1{,}65 \cdot u_c \tag{138}$$

where u_c is the combined uncertainty of the test result

4.6. Reporting the test result

The test process yields a value as the estimate for the conventional true value of the measurand. In principle, this value is the sample mean $\bar{y}$ or simply y. As standards strongly recommend [1, 2, 4], the y value must be reported together with its expanded (extended) uncertainty U for a specific confidence level (typically 95%) as follows:

$$y \pm U \tag{139}$$

The laboratory may specify the *pdf* type and k value as substitutes for the confidence level specification. If in a testing situation a laboratory could not evaluate a metrological sound numerical value for each component of *MU* then this laboratory may report the standard uncertainty of repeatability, but this shall be clearly stated as it is recommended in [3, 14].

The laboratory practice shows that *MU* estimation is in many cases a time-consuming and costly task. This endeavor shall be justified by the advantage of *MU* evaluation for testing laboratories. EA 4/16 argues that *MU* assists in a quantitative manner important issues such as risk control, credibility of the test results, competitiveness by adding value and meaning to the test result etc. But, in the author's opinion, the *MU* estimation imposes the operator increased awareness about the influence factors and higher interest in the means to improve the quality of the test results.

5. Uncertainty estimation for Rockwell C hardness test. Case study

5.1. Hardness testing-general principles

Hardness is the measure of how resistant a solid bulk material is against ingression by spherical, pyramid or conical tips. Hardness depends on the mechanical properties of the sample as: ductility, elasticity, stiffness, plasticity, strain, strength, toughness etc [15]. On the other hand, quantity "hardness" cannot be expressed as a numeric function of a combination of some of these properties. Therefore, hardness is a good example of measurand that cannot be mathematically modeled without referring to a method of measurement [4, 15]. In this view, there is a large number of hardness testing methods available (e.g. Vickers, Brinell, Rockwell, Meyer and Leeb) [15, 16]. As a consequence, there is not a measure unit for hardness independent of its measurement method. In time, based mainly on empirical data, standard methods for hardness testing appropriate to the material grades were developed. In the metallurgical testing field the Brinell, Rockwell and Vickers methods are the most frequently used. For this reason, this case study addresses the well-known Rockwell C hardness by measuring the depth of penetration of an indenter under a specific load. There are different Rockwell type tests denotes with capital letter A÷H [16, 17], but herein referred only C type.

5.2. Description of the measurand

The Rockwell C hardness is the measure of how resistant a solid bulk material is against penetration by a conical diamond indenter, having a tip angle of 120°, which impinges on the flat surface of the sample with a prescribed force.

The Rockwell C hardness scale is defined as:

$$h_{RC} = 100\text{x}(0.002) - d \tag{140}$$

where d is the penetration depth.

The most meaningful measure for metallurgists is the Rockwell C index, defined as:

$$HRC = h_{RC}/0{,}002 \tag{141}$$

5.3. Description of the test method

The method used for Rockwell C is given in ISO 6805 standard [16]. The Rockwell C test is performed on bulk sample having geometrical dimensions complying with the ISO 6805 specifications. The indenter is diamond cone having the tip angle of $120^0 \pm 0.5^0$. The tip ends with a hemisphere of 0.002 mm in diameter. The indentation is a stepwise process i.e. in the first step the indenter is brought contact with the sample surface and loaded with a preliminary test forces F_0=98 N with a dwell time of 1-3 s. Subsequently, during a period of 1s - 8 s, an additional force F_1= 1373 N is loaded. The resulting force F=1471 N is applied for 4±2s. In the third step the F_1 force is released and the penetration depth h is measured in the period 3s-5s. The ISO 6805 recommends that the number of tests carried on in repetitive (reproductive) conditions to be a multiple of 5. In this instance, 5 indentations are established as practice for estimating the conventional true value of HRC of the specimen that has undergone the test. The recommended environmental conditions of the test are [16]: temperature 10-35°C, protection against: vibrations, magnetic and electric fields, avoiding the soiling of the sample.

5.4. The mathematical model of the measurand

The mathematical model given by GUM for estimating *MU* of the hardness reference blocks is the most appropriate for the *MU* estimation in the industrial hardness testing. The mathematical model given by GUM takes into account the correction factors coming from equipment calibration as follows:

$$h_{RC} = f(\bar{d}; \Delta c; \Delta b; \Delta e) = 100 \times 0.002 - \bar{d} - \Delta_c - \Delta_b - \Delta_e\ (mm) \quad (142)$$

where $\bar{d}$ is the mean of the penetration depth; Δ_c is the correction assigned to the equipment ; Δ_b is the difference between the hardness of the areas indented by reference equipment and the calibrated area; Δ_e is the correction assigned to the uncertainty of the reference equipment.

Δ_c is assigned to the equipment by comparing its penetration depths, done in a primary hardness reference block, with those of a reference equipment („national etalon machine").

Although Δ_b and Δ_e are negligible corrections, they are introduced in the expression of the mathematical model to be taken into account as contributors to the *MU* budget. According to ISO 6508-1:2005 [16] a „Hardness Reference Block" may be considered as CRM (Certified Reference Materials). Based on the above model it was considered that the h_{RCX} of a X specimen may be expressed as:

$$h_{RCX} = f(\bar{d}; \Delta c; \Delta b; \Delta e) = 100 \times 0.002 - \bar{d}_X - \Delta_C\ (mm) \quad (143)$$

where $\bar{d}_X$ is the mean of the penetration depths given by the testing machine on X specimen; Δ_c is the correction assigned to the equipment resulted by comparring the mean penetration depth done on a CRM with those specified by its certificate i.e.

$$\Delta_C = d_{oCRM} - \bar{d}_{CRM} \quad (144)$$

where $\bar{d}_{CRM}$ is the mean of the penetration depth given by the test machine on CRM, d_{oCRM} is the expected penetration depth derived as:

$$d_{oCRM} = (100 - HRC_{CRM}) * 0.002 \quad (145)$$

where $H_{RC\text{-}CRM}$ is the certified Rockwell C hardness of the CRM

5.5. Estimation of the contribution of the influence factors to the overall *MU*

5.5.1. *Contribution assigned to the mean indented depth on specimen* ($\bar{d}_X$)

The contribution of $\bar{d}_X$ to the averall *MU* is assessed by the standard deviation of the mean which is a standard uncertainty (type A). In addition to the data spreading around $\bar{d}_X$, the uncertainty of the resolution of depth measuring system will be considered. The contribution of $\bar{d}_X$ to the overall *MU* is:

$$u_c^2(\bar{d}_X) = t^2\frac{S_X^2}{n} + \frac{\delta^2}{12} \tag{146}$$

where S_X^2 is the sample variance of the identation depths, n is the number of indentations, t is the Student factor, $\delta^2/12$ is uncertainty assigned to depth reading having a triangle *pdf* [2]

The S_X^2 can be calculated as:

$$S_X^2 = \frac{1}{n-1}\sum_{i=1}^{n}(d_i - \bar{d}_X)^2 \tag{147}$$

where d_i are the indentation depth, $i = \overline{1,n}$., n is the number of indentations

5.5.2. *Contribution assigned* Δ_c

The uncertainty assigned to Δ_c has two contributors i.e. the standard uncertainty assigned to the measurement process on CRM and that assigned to the CRM itself, which is of B type. The *MU* attributed to $\bar{d}_{CRM}$ is of the same type with that assigned to $\bar{d}_X$ therefore it may be expressed as:

$$u^2(\bar{d}_{CRM}) = \frac{t^2 S_{CRM}^2}{n} + \frac{\delta^2}{12} \tag{148}$$

where S_{CRM}^2 is the sample variance of the depths given by indentations carried on the CRM, $\delta^2/12$ has the same significance as in Eq.(7), t is the Student factor.

S_{CRM}^2 can be calculated using the Eq.(147)

The uncertainty assigned to the CRM is taken from its certificate for *k*=1 and is denoted as u_{MRC}. In this instance, the combined uncertainty assigned to Δ_c is:

$$u_c^2(\Delta_c) = \frac{t^2 S_{MRC}^2}{n} + \frac{\delta^2}{12} + u_{MRC}^2 \tag{149}$$

The common practice of the *MU* assessment is to take into account the uncertainties assigned to machine calibration and that attributed to the operator. But, these contributions are included in the *MU* assigned to $\bar{d}_X$ and to Δ_c. Thus, the supplementary addition of the mentioned *MU* will involve an overestimation of combined uncertainty of the hardness test.

5.6. Combined uncertainty calculation

It is a matter of evidence that the contributions to the overall uncertainty of Rockwell C hardness test are mutually independent and, consequently, the combined uncertainty u_c assigned to h is:

$$u_c(h_{RC}) = [u_c^2(\bar{d}_X) + u_c^2(\Delta_c)]^{1/2} = t^2(\frac{s_X^2}{n} + \frac{s_{MRC}^2}{n}) + \frac{\delta^2}{6} + u_{MRC}^2 \quad (150)$$

5.7. Extended uncertainty calculation

The reference documents such as [2, 17] do not specify a calculation methods for extended *MU* of h_{RC}. These reference documents only specify the combined uncertainty of the measurement. ISO 6508 gives clear clues that a Gaussian *pdf* is assigned to $\bar{d}_X$. These are: *t* factor e.g. *t*=1.14 for *n*=5, and *k*=2 for U_{CRM} (Eq.(B.6) in [16]). Underpinned on ISO 6805 one can assign a Gaussian *pdf* to $\bar{d}_X$. Thus, the extended uncertainty can be calculated by multiplying $u_c(h_{RC})$ with a coverage factor *k*, for instance, *k*=2 for a confidence level of 95%. Thus, the expanded uncertainty assigned to h_{RC} is:

$$U = U(h_{RC}) = k(u_c(h_{RC})) \quad (151)$$

According to metallurgical practice, the expanded *MU* of the test result is:

$$U(95)2(u_c(h_{RC})) \quad (152)$$

Thus, the expanded uncertainty of the HRC index is calculated as:

$$UHRC = U(h_{RC})/0.002 \quad (153)$$

5.8. Reporting the result

If the bias of the test is corrected then the result in HRC unit is:

$$HRCcorr = 100 - (\bar{d}_X + \Delta_c)/0.002 \quad (154)$$

In this case, the result shall be reported as:

$$HRC_{corr} \pm U \quad (155)$$

on condition that the level of confidence of *U* or the *k* value shall be specified.

If the *HRC* value is not corrected then the *HRC* index is:

$$HRC = 100 - (\bar{d}_X/0.002) \quad (156)$$

and the assigned expanded *MU* to the test result is estimated as:

$$U + \Delta_c/0.002 \quad (157)$$

In this case, the result shall be reported as:

$$HRC \pm (U + \Delta_c / 0.002) \tag{158}$$

5.9. Numerical example

5.9.1. General data about the Rockwell C hardness test

A Rockwell C hardness test was performed in reproductive conditions on a steel specimen using a Balanta Sibiu Rockwell Machine indirect calibrated on a CRM i.e. a hardness reference block having a certified hardness of 40.1 *HRC* with a *U(95)* = 0.22 HRC. The calibration data obtained on CRM SN 48126 are given in Table 1.

No.	1	2	3	4	5	Mean	SD	SDmean
d_i	0,1202	0,12	0,1192	0,1191	0,1189	0,12	0,00058	0,00026
h_i	0,0798	0,08	0,0808	0,0809	0,0811	0,08	0,00058	0,00026
HRC_i	39,90	40,00	40,40	40,45	40,55	40,26	0,290	0,130

Table 1. The calibration data obtained on CRM SN 48126

Subsequently, according to [16] 5 indentations were carried on the specimen, denoted *X*. The hardness test data for the *X* specimen are given in Table 2.

No.	1	2	3	4	5	Mean	SD	SD_{mean}
d_i	0,1166	0,1148	0,115	0,1154	0,117	0,12	0,00098	0,00044
h_i	0,0834	0,0852	0,085	0,0846	0,083	0,08	0,00098	0,00044
HRC_i	41,7	42,6	42,5	42,3	41,5	42,12	0,49	0,22

Table 2. Test data for the *X* specimen

From Table 1 and Table 2 it results Δc= - 0.15 HRC.

5.9.2. MU estimation for the test result according to the classical approach

The u_c assigned to the test result is estimated based on data provided by: testing, calibration procedure, certificate of the CRM and on the specification of operating manual. The data used for u_c calculation are given in Table 3.

No.	Uncertainty source	Symbol	Uncertainty[HRC]	Evaluation type	Assigned pdf
1	Indirect calibration	$s_{M\text{-}MRC}$	0.15	A	Gaussian[1]
2	MRC	U_{MRC}	0.11	B	Triangular[1]
3	Sample	$s_{M\text{-}X}$	0,25	A	Gaussian[1]
4	Displaying resolution	δ	0.05	B	Triangular[1]

Table 3. The input data used for u_c calculation. 1-the assigning of *pdfs* are taken after [16] but there are no evidences for these assignments.

The combined uncertainty calculated by replacing the values given in Table 3 in Eq.(147) is:

$$u_c = 0.31\ HRC \tag{159}$$

According to [16] the corrected HRC (see Eq(154)) is:

$$HRC_{corr} = 41.96 \pm 0.62\ HRC \tag{160}$$

where $U = 0.62\ HRC$ is the expanded uncertainty for the 95% confidence level calculated with a coverage factor $k = 2$ according to Anex B of ISO 6805-2.

5.9.3. MU for the test result according to the multiconvolutional approach

Whatever is the way of reporting the results, it is quite impossible to assign a confidence level to the expanded uncertainty of the result based on irrefutable arguments. The alternative approach is to consider that the measurand has a uniform *pdf*. The worst case is a uniform *pdf* having a width equal to the interval of obtained values i.e.

$$f_{d(x)} = \begin{cases} \frac{1}{d_{max}-d_{min}}, & x \in [d_{min} - U_{cc}; d_{max} + U_{cc}] \\ 0, otherwise \end{cases} \tag{161}$$

where d_{max} and d_{min} are the maximum and the minimum penetration depths among the five test data. The mean of $f_{d(x)}$ is $\mu = (d_{max} + d_{min})/2$ while the half width of the rectangular distribution is $= (d_{max} - d_{min})/2$. According to the data given in Table 2, $\mu = 0.1159\ mm$, $\Delta_c = 0.003$ mm and $a = 0.0125$ mm. As shown in &3.2.3 the probability that the mean m of a 5 reproductive outcomes, each one being uniformly distributed in [$-a$, a], to belong to the interval $[m - a/10; m + a/10]$ is 29.4%. while in $[m - a/5; m + a/5]$ is 57.64% Thus, the conventional true value of the penetration depth lies in the interval [0.1157; 0.1161] mm with a confidence level of 55%. In this instance, the *HRC* index of the specimen could be reported as:

$$HRC\text{corr} = 41.89 \pm 0.55\ HRC \tag{162}$$

for the 55% confidence level calculated on the basis of *pdf* of mean.

The confidence level for $\left|\frac{m-\mu}{a}\right| \leq \frac{2}{5}$ is about 99%. In this case, Rockwell C index could be reported as:

$$HRC\text{corr} = 41.89 \pm 1.1\ HRC \tag{163}$$

The above result should be interpreted as the *HRC* conventional true value of the specimen lies in the interval [40.79, 42.99] (HRC) with 99% confidence level. For the conventional case adopted by ISO 6805 and presented above, the 99% confidence level corresponds to a coverage factor *k*=3 i.e. to a *U*(99%)=0.93 *HRC*. By comparison, the reported test result, according to [16], as $HRC_{corr} = 41.96 \pm 0.93\ HRC$ with 99% confidence level or, according multiconvolutional approach, as $HRC_{corr} = 41.89 \pm 1.10\ HRC$ seems to be quite the same to some extent depending on the rigor claimed by the client. But, from scientific and mathematical statistics point of view one may feel comfortable to use a founded test result having assigned a little bit larger uncertainty than to use a doubtful one.

5.10. Discussion

The most important finding of this case study is that, when dealing with a measurand whose assigned *pdf* is not known or is insufficiently documented, the best approach is to consider it has a uniform distribution. The interval of variance of the outcomes may be considered, at first glance, as the distribution width. Another important issue is that for estimating *MU* using the multiconvolutional approach, only data provided by the testing process are used, while the classical approach uses supplementary data. A sound question regarding the multiconvolutional approach is how to decrease the *MU* i.e. to improve the test result quality? The classical solution is that the experimentalist should increase the number of reproductive or repetitive tests. The common practice of five tests appears insufficient, but ten tests should be acceptable. Using a ten-fold convolved uniform *pdfs* one can describe quite accurately the probability of mean displacement about the conventional true value (μ) with at least 0.1μ increment.

6. Summaries

This article addresses a more meaningful approach for measurement uncertainty estimation, particularly in the metallurgical test field. The chapter contributes to the state of the art by the development of a consistent approach for calculating the *pdfs* of the sample mean statistic, of the variance and of *t* parameter. To this end, the concepts of probabilistic theory and the derivation of the main *pdfs* used for measurement uncertainty evaluation as Poisson, Gauss are presented briefly. The theoretical backgrounds are presented in the paper in the aim to make clear for an experimentalist the specific *pdfs* to be assigned to a measurand. A considerable part of the chapter addresses the deriving of the *pdfs* of the compound variables as $Y=aX+b$, $Y=(X_1+....+X_n)/n$, $Y=X^2$, $Y=\sqrt{X}$, $Y=X_1/X_2$ because these form the basis for deriving the probability density functions of sample mean, of variance, of Student and of Fisher-Sedecor. The *pdfs* of sample mean and of Chi distribution are derived *in extenso* following the convolution approach because it provides an easy-to-use and intuitive way to understand how these distributions should be applied for measurement uncertainty estimation. Thus, this approach allows an in-depth understanding of the mathematical formulas in order to avoid their usage exclusively based upon the mathematical literature without understanding of or without concern about the appropriateness to the case addressed. An important contribution of the chapter is the argument for using a number of repeated test, *n*, as the number of degrees of freedom and not *n*-1 as is common practice. The last part of the chapter deals with measurement uncertainty estimation using GUM method because it is required by EN ISO/CEI 17025. The GUM approach was applied for the uncertainty estimation of the Rockwell C hardness test according to the ISO 6825 standard. As is underlined in the chapter, this standard does not provide clear evidence for assigning Gaussian distribution to the hardness HRC. Alternative approach to estimate the uncertainty of the Rockwell C hardness test result is given based on the *pdf* of the sample mean obtained by 5-fold convolved product of the uniform distribution assigned to the measurand.

The entire chapter is designed to emphasize the risk of wrongly estimating the test result uncertainty due to erroneous assumptions regarding the *pdf* attributed to the measurand or to the influence factors of the uncertainty budget.

Author details

Ion Pencea
Metallic Material's Science and Engineering Department, Materials Science and Engineering Faculty, University "Politehnica" of Bucharest, Bucharest, Romania

7. References

[1] EN ISO/IEC 17025:2005, General requirements for the competence of testing and calibration laboratories
[2] Guide to the Expression of Uncertainty in Measurement, Edition, ISO, Geneva, 1995,
[3] Sfat, C.E, (2011) Quality Assurance of Test Results, in: Saban, R., editor, Treatise of Science and Engineering of Metallic Materials, AGIR Publisher, Bucharest, Romania, pp 1272-1287.
[4] Eurolab Technical Report 1/2007-Measurement Uncertainty Revised, pp 47-50
[5] Cuculescu, I., (1998) Probability Theory, All Publisher, Bucharest, Romania, 510p.
[6] Ventsel, H., (1973) Theorie des Probabilities, MIR Publisher, Moscou, 565p
[7] Dehoff, R.T., Rhines, F.N., (1968) Quantitative Microscopy, McGrow-Hill Book Company, 421p.
[8] ISO 3534-1:2006, Statistics -Vocabulary and symbols - Part 1: General statistical terms and terms used in probability
[9] Renyi, A., (1970) Probability Theory. North-Holland, Amsterdam, 458p.
[10] Frank Killmann, F., von Collani, E., (2001) A Note on the Convolution of the Uniform and Related Distributions and Their Use in Quality Control, Heldermann Verlag Economic Quality Control, Vol 16 , No. 1, pp.17 – 41
[11] ISO 5725 (6 parts), Accuracy (trueness and precision) of measurement methods and results
[12] Pencea, I., (2011) Methods and Techniques of Instrumental Analysis of Materials in: Saban, R., editor, Treatise of Science and Engineering of Metallic Materials, AGIR Publisher, Bucharest, Romania, pp 1057-1234.
[13] VIM-International Vocabulary of Metrology – Basic and General Concepts and Associated Terms, JCGM 200:2012,
[14] EA Guideline EA-4/16: Expression of Uncertainty in Quantitative Testing EA 2003, Guidelines for the estimation of uncertainty in environmental measurement, (www.european-accreditation.org)
[15] Popescu, N., Bunea, D., Saban, R., Pencea, I., (1999) Material Science for Mechanical Engineering, v1, Fair Partner Publisher, Bucharest, Romania, 270p.
[16] ISO 6508-1:2005, Metallic materials -- Rockwell hardness test -- Part 1: Test method (scales A, B, C, D, E, F, G, H, K, N, T)
[17] EA Guideline EA-10/16: EA Guidelines on the Estimation of Uncertainty in Hardness Measurement, (www.european-accreditation.org),

Chapter 7

Artificial Intelligence Techniques for Modelling of Temperature in the Metal Cutting Process

Dejan Tanikić and Vladimir Despotović

Additional information is available at the end of the chapter

http://dx.doi.org/10.5772/47850

1. Introduction

The heat generation in the cutting zone occurs as a result of the work done in metal cutting process, which is consumed in plastic deformation of the cutting layer and overcoming of friction, that occurs on the contact area of the cutting tool (i.e. cutting insert) and work material (i.e. workpiece). The heat generated in the chip forming zone directly influences the quality and accuracy of the machined surface. The negative occurrences in the metal cutting process, such as: Built Up Edge (BUE) formation, work-hardening, plastic deformation of the cutting edge, deformation of the workpiece, etc. are also dependent on the heat.

Modelling of temperature in the metal cutting process is very important step in understanding and analysis of the metal cutting process. In order to model the temperature which occurs in the chip forming zone, large number of experiments must be carried out at different cutting conditions, synchronously measuring the chip's top temperature using the infrared camera. The infrared method gives a relatively good indication of the measured temperature, comparing with other methods for temperature measurement, such as: thermocouples, radiation methods, metallographic methods etc.

In recent years the research in the area of process modelling is directed on the use of systems based on artificial intelligence: artificial neural networks, fuzzy logic systems, genetic algorithms, as well as combination of mentioned systems. Results obtained in the first phase will be used for modelling of the cutting temperature using the response surface methodology model (RSM model), feed forward artificial neural networks (ANN model), radial basis function network (RBFN model), generalized regression neural network (GRNN model) and adaptive neuro-fuzzy system (NF model). The accuracy of the proposed models will be presented, as well as their suitability for use in concrete problems.

Analysis and modelling of the metal cutting process can be very useful in determining of the optimal values of input process parameters (cutting speed, depth of cut and feed rate). Positive effects could be many. The quality of the machined surface can be enhanced and tool life can be extended, leading to advancement of the production economy.

2. Metal cutting process temperature

Cutting temperature affects changes in the workpiece material, and consequently, the quality of the machined surface. It also influences changes in cutting tool material and plays an important role in tool wear. Chip temperature might be used to investigate the friction behaviour of cutting tools, because this temperature is dependent on the friction energy which is entering the chip at the rake face.

The amount of the heat generated in the metal cutting process is expressed through the work done in the process and the mechanical equivalent of the heat (Arshinov & Alekseev, 1979), in the form:

$$Q = \frac{F_z v}{E} \tag{1}$$

where: Q – amount of the heat generated in the metal cutting process, $F_z v$ – work done in the process, E – mechanical equivalent of heat

The heat balance during the metal cutting process can be expressed as follows:

$$Q = Q_1 + Q_2 + Q_3 + Q_4 \tag{2}$$

where: Q – total amount of heat generated in cutting, Q_1 – amount of heat carried away in the chips, Q_2 – amount of heat remaining in the cutting tool, Q_3 – amount of heat passing into the workpiece, Q_4 – amount of heat radiated to the surrounding air

According to the empirical investigations, 60-86% of the heat is carried away in the chips and grows with increase in cutting speed. For lathe operations this proportion is as follows: 50-86% of the heat is removed in the chip, 10-40% remains in the cutting tool, 3-9% left in the workpiece and about 1% radiates into the surrounding air.

A large number of factors affect the quantity of heat generated. The most important ones are: the cutting speed and the cutting depth (Tanikić et al., 2010a). It is also noticed that there is more heat transferred into the workpiece in the finishing turning than in the rough turning. Theoretically, there are three zones of the heat generation that can be identified during turning (Fig. 1.) (Tanikić et al., 2010b):

- cutting zone
- tool-chip contact zone
- tool-workpiece contact zone

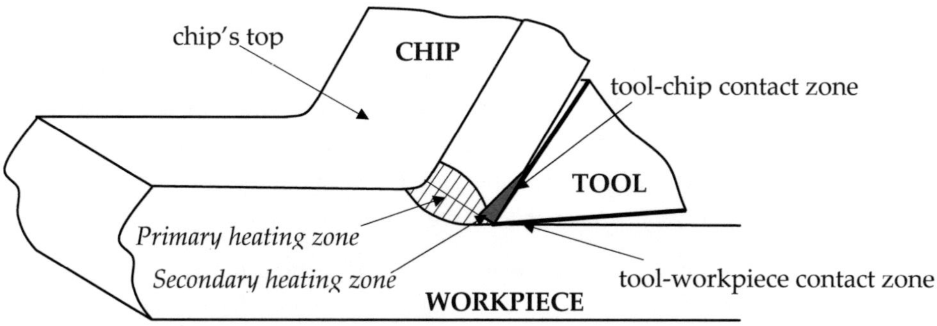

Figure 1. Heat generation zones during metal cutting process

The temperature of the various points of cutting tool, chip and workpiece are different, as shown in Fig. 2. (Arshinov & Alekseev, 1979). Temperature of the layers close to the cutting tool surface is higher than those away from it. The highest temperature, as expected, occurs at the point of cutting tool – workpiece contact (denoted with T on Fig. 2.), while the temperature of the other points are given as proportion of this temperature.

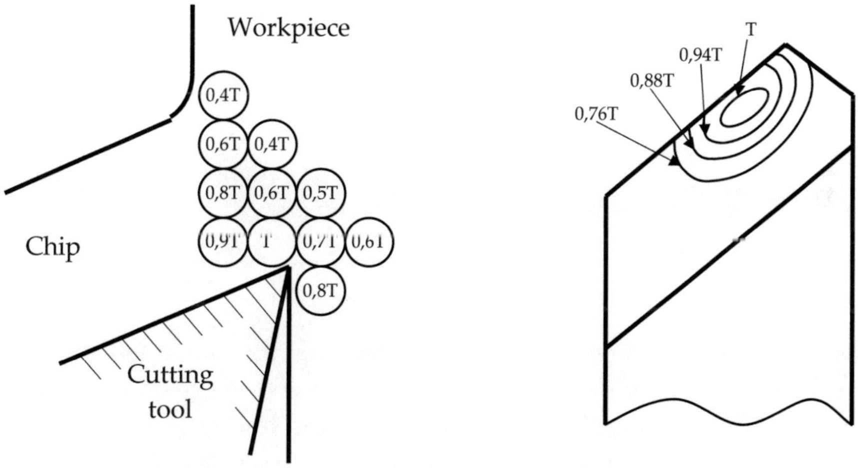

Figure 2. Temperature distribution in the cutting tool, chip and workpiece

2.1. Factors influencing cutting temperature

Factors which directly influence cutting temperature, as well as chip temperature, during metal cutting process are: type of workpiece material, cutting regimes (cutting speed, feed rate and depth of cut), dimensions and geometric characteristics of cutting tool, quantity and pressure of the coolant fluid etc. Recent investigations show that bar diameter also influences cutting temperature (Boud, 2007).

Workpiece material – In general, the greater amount of heat is generated during cutting steels when compared to cast iron. Cast irons also have lower thermal conductivity than steels. The high pressure on the peak of cutting tool during cutting cast iron causes short and broken chips. Mechanical properties of the workpiece material affect cutting temperature significantly. The higher the tensile strength and hardness of the workpiece, the greater the resistance force that must be overcomed during cutting, i.e. the more work is required to be done, resulting in higher cutting temperatures. On the other hand, the higher thermal conductivity and heat capacity of the workpiece, the higher the level of the heat transfered from the place where it is generated into the chip and the workpiece. At the same time there is lower temperature in the surface layers of the tool.

Cutting regimes – The cutting forces disproportionally decrease with an increase in the cutting speed (Tanikić et al., 2009a). For example, increasing in cutting speed by approximately 500% causes decreasing in the cutting force of about 21% (O'sullivan & Cotterell, 2001). The amount of the heat generated during the metal cutting process depends on both factors: the cutting speed and the cutting force. Generally, the higher temperature is generated with increasing the cutting speed. The cutting force, as stated previously, increases disproportionally with an increase in the feed rate and therefore the cutting temperature increases, too. Increase of the cutting force and the cutting temperature is slower than the feed rate increase. Results of many experiments show that the cutting temperature depends on a large number of factors, which can be expressed by the following equation (Radovanović, 2002):

$$T = C_T a^{k_{aT}} f^{k_{fT}} v^{k_{vT}} \tag{3}$$

where: T [K] – cutting temperature, a [mm] – depth of cut, f [mm/rev] – feed rate, v[m/min] – cutting speed, C_T – general coefficient, k_{aT}, k_{fT}, k_{vT} – exponents

General coefficient C_T and exponents: k_{aT}, k_{fT}, k_{vT} depend on the workpiece and the tool material characteristics, tool geometry, type of coolant etc...

Tool geometry – Cutting temperature directly depends on the cutting tool angles as well as the nose radius. The cutting tool angles define the size and the position of the maximum heated area. The larger the nose radius, the greater the resistance force and the cutting temperature. Increasing the nose radius also has a positive effect, such as increasing the active cutting edge, i.e. the area which is in the focus of deformation. In that way, better heat conduction through the tool and the workpiece is provided.

Type of coolant – Using the coolant fluids the temperature is reduced in two ways. Firstly, an amount of generated heat is carried away directly with the cutting fluid. The second one is the positive influence on the lubrication and the reduction of the friction between the tool and the workpiece. The coolant fluid jet must be directed to the contact point of the tool and the workpiece, while the quantity of the used fluid depends on the cutting speed.

2.2. Methods for temperature measuring in metal cutting process

A large number of temperature measurement methods in the metal cutting have been developed in the past years. This section gives a brief history of these methods.

Thermocouples – Thermocouples are frequently used transducers in temperature measuring because they are rugged, they cover a wide temperature range and they are relatively inexpensive (O'sullivan & Cotterell, 2001). When two dissimilar metals touch each other, the contact point produces a small open circuit voltage, which is proportional to the temperature difference of the connected metals. If these two metals are the tool and the workpiece, this thermocouple is then called a "tool-work thermocouple", or "tool-chip" thermocouple. These kind of thermocouples are used for temperature measurements in the contact area of the tool and the chip. The cutting zone forms, so called, a "hot junction", which produces thermo-electric emf (electro magnetic force), while the cutting tool and/or the workpiece form, so called, a "cold junction". This technique is usually used for measuring the average temperature of the whole contact area. It is almost impossible to measure any brief variations of temperature with this method. An error occurs in temperature measurement when a BUE is formed during the cutting process. A drawback of this method is the fact that a coolant fluid cannot be used during the measurement. The cutting tool and the workpiece must be built from an electro conductible material and the system calibration is necessary on every single setup. The constraint of this method is also in the type of workpiece material, which can't be made from an easily melted material.

Inserted thermocouples – In order to improve the performance of the earlier mentioned system, thermocouples are inserted into the cutting tools in a special way which allows them to measure the temperature in a single or several points at the same time (Childs et al., 2000). The negative side of this method is that it requires drilling a few holes in the tool or in the workpiece, where the thermocouples are nested, very closely to the place where temperature is measured, in order to ensure accuracy. This method was used for measuring cutting temperatures in cutting steels, and cutting various alloys, on the lathes and on the milling machines (Kitagawa et al., 1997).

Radiation methods – This category includes methods for measuring temperature at a single point, or measuring temperature field, without the direct contact between the measuring instrument and the object. In the single point temperature measurements an infrared pyrometer is used, while in the measuring temperature field (infrared thermograph) specially made infrared cameras sensitive to radiation of the body which is heated are used (Abukhshim et al., 2006). Radiation methods have a large number of conveniences with respect to conduction methods. The most important are: faster response of the system, i.e. possibility of measuring brief variations of the temperature, there is no negative influence on the tool and the work material, there is no physical contact between the measuring system and the object, remote temperature measuring for inaccessible objects etc... During temperature measurement process using the infrared camera, there is a possibility that unwanted hiding of the measuring point with the chip may happen, which implies obtaining faulty data (Young, 1996). The other negative side of this method of temperature measurement is the fact that it requires knowledge of the exact value of the coefficient of emission for precise measuring. In order to overcome this problem, the area of interest can be painted with the paint with known coefficient of emission. Coefficient of emission depends on the clarity of the target area, presence of the oxidation covering, the wave length etc. Any of the above mentioned factors have an influence on the distraction of measured data.

Metallographic techniques – This method involves analysis of microstructure and/or micro hardness of the heat affected zones. It requires calibration curves which show the level of dependence of the material hardness in terms of the known temperatures and the time of heating. The usual accuracy of this method is ±25° (Wright, 1978). These methods are mainly used for temperature measurements of the cutting tools made from high speed steels, because they show structural changes, and/or hardness, in the temperature range of 600-1000 °C.

Other methods – These methods include methods for temperature measurement using thermo-sensitive paints, liquid crystals, fine powders etc. and are mainly limited to measurement of the visible areas in the special laboratory conditions (Ay & Yang, 1998).

Generally, there are no strictly defined rules to determine which method is the most adequate one in a given situation. On the other hand, high complexity of the process itself does not always permit to compare results obtained by different methods. The best illustration of the above mentioned is the fact that even the results obtained by the same method in completely identical experiment conditions can be different, which is another proof of complexity of temperature measurement in the metal cutting process.

3. Experimental research

The lathe, which was used for examining and measuring, is located in the Laboratory for Production Engineering, at the Mechanical Engineering Faculty in Niš, Serbia. The workpiece material used is steel, with AISI designation 4140. This steel belongs to the group of doped, decent, cold drown steels, with strength of R_m=1050 [N/mm^2]. Four thermally treated metal specimens, with measured hardness of HRC 20, 36, 43 and 55, were machined. The dimensions of the workpieces are ϕ45x250 [mm].

Fig. 3. shows the component relations and information flow of the material handling system, and linked information system, which processes the obtained data.

SANDVIK Coromant cutting tool has been chosen, which consists of two parts: tool holder PCLNR 32 25 P12 in combination with cutting insert CNMG 12 04 08 (grade 235), according to the recommendations of the manufacturer and the empirical knowledge.

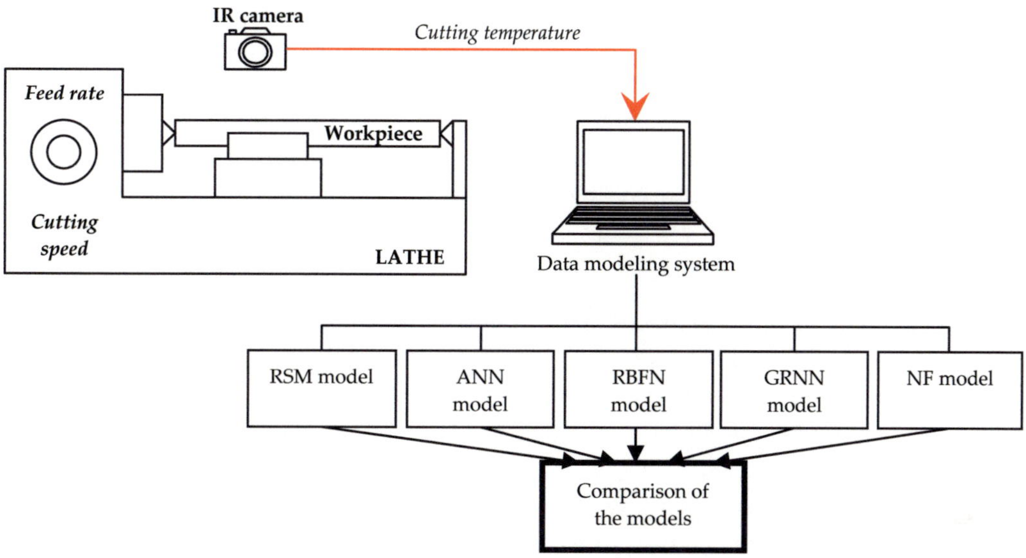

Figure 3. Schematic representation of the information flow in the system

Jenoptik Varioscan 3021-ST infrared camera has been used for temperature measurement. Varioscan high resolution is scanning thermovisics measured system, for wave lengths outside of the vision spectrum, from 8μm to 12μm, i. e. in the area of infrared emission. Signal from this spectrum is amplified, digitalized with 16 bites and visualized. Every color on the shown thermagram (Fig. 4.) represents particular temperature. Temperature resolution of this system is 0.03°C, while operating range of the camera is -40°C to +1200°C.

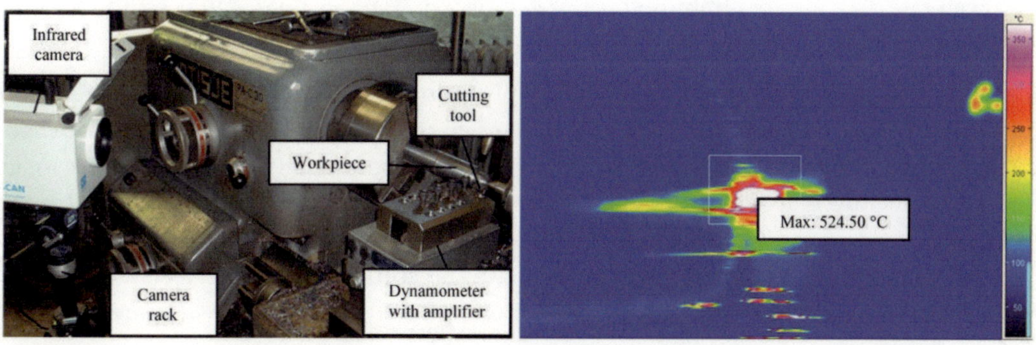

Figure 4. Experiment setup and thermagram (v=125[m/min], f=0.196[mm/rev] and t=2[mm])

The most important temperature, from the metal cutting process point of view, is maximum temperature of the cutting tool. This temperature directly affects cutting characteristics of the tool, tool and workpiece deformation as well as the quality of the machined surface. It is obvious that measuring of the rake face of the cutting insert, where maximum temperature occurs, is not possible using mentioned infrared camera, because of continual presence of the chip which covers the area of interest. With known values of chip's top temperature, cutting depth and physical properties of the workpiece it is possible (using, for example, finite-difference model or FEM analyses) to calculate maximum cutting tool temperature. However, the primary goal of this work is exploring the possibility of using the means of artificial intelligence in modelling cutting temperature (and not measuring the exact value of maximum cutting temperature), and that's the reason why chip's top temperature is adopted as relevant parameter.

In the beginning of the metal cutting process temperature rises, until it reaches the maximum value. That's the reason why the measurement should be performed a small period of time after the beginning of the process (Kwon et al., 2001). After observing rise in the temperature and its distribution at the beginning of the process with an infrared camera, it is concluded that a period of about 60 seconds is enough for stabilizing the measured temperature. The pictures (Fig. 4.) are submitted to a PC memory card, and later analysed. The maximum cutting temperature which occurs on the chip's top is used for temperature modelling in the next phase of this work.

3.1. Measured results and discussion

Modelling of the chip's top temperature requires a large number of experiments with different cutting regimes. As mentioned, in addition to the recommended data obtained from the appropriate literature, the empirical knowledge is of crucial importance in making a proper choice of cutting regimes. The adopted variable process parameters are:

- Material's hardness HRC (values: HRC 20; 36; 43 and 55)
- Cutting speed v[m/min] (values: 80; 95; 110; 125 and 140 [m/min])
- Feed rate f [mm/rev] (values: 0.071; 0.098; 0.196 and 0.321 [mm/rev])
- Depth of cut t [mm] (values: 0.5; 1; 1.5 and 2 [mm])

Results of the temperature measurement are given in Fig. 5.a. to 5.d. From the presented figures, it can be concluded that, with increasing the cutting speed (while other parameters remain constant) the resistance force of the cutting increases too, resulting in the increase in the chip temperature. It is also obvious that larger values of temperature occur in machining of the hardened workpiece materials. The chip temperature increases with increasing the depth of cut, too (with constant values of feed rate and cutting speed). The feed rate also has influence on changes in the cutting temperature, which is particularly apparent at low cutting speeds.

The irregularities in the following figures (the peaks in the diagrams of the chip's top temperature) can be interpreted as measuring errors. Anyway, the correlations among cutting regimes and corresponding temperature (trendlines of chip's top temperature) can be achieved.

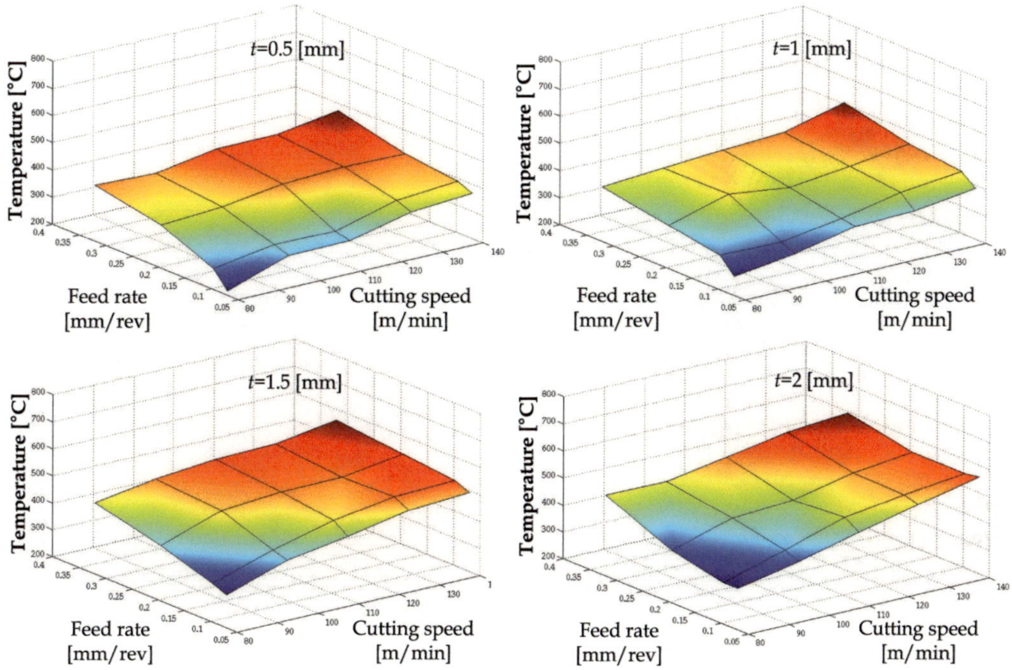
t=0.5 [mm]
t=1 [mm]
t=1.5 [mm]
t=2 [mm]
Temperature [°C]
Feed rate [mm/rev]
Cutting speed [m/min]

(a)

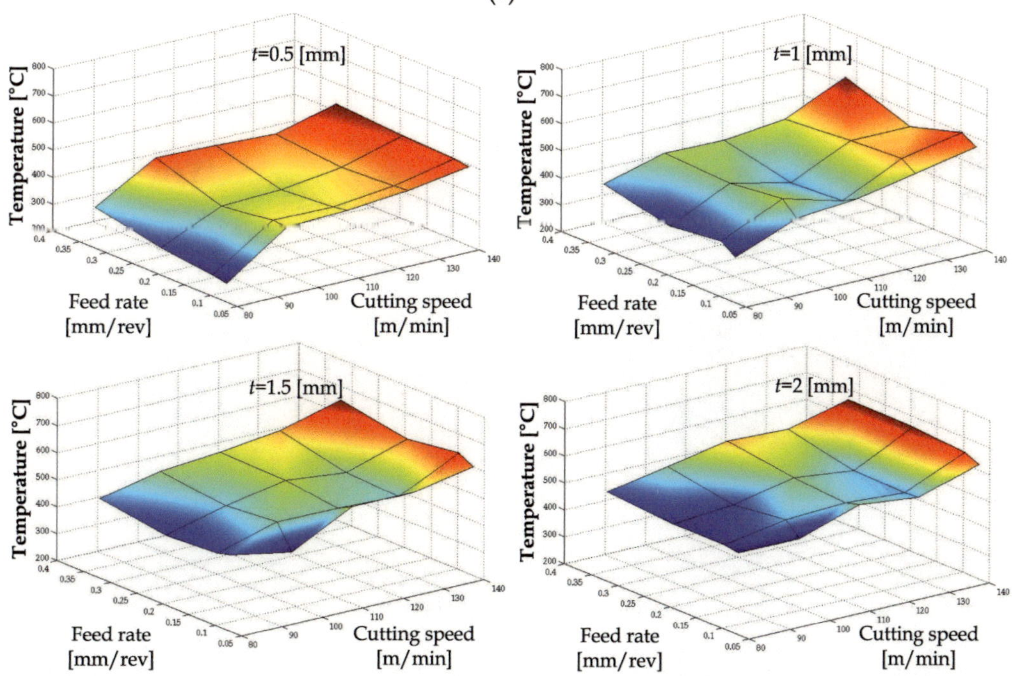
t=0.5 [mm]
t=1 [mm]
t=1.5 [mm]
t=2 [mm]
Temperature [°C]
Feed rate [mm/rev]
Cutting speed [m/min]

(b)

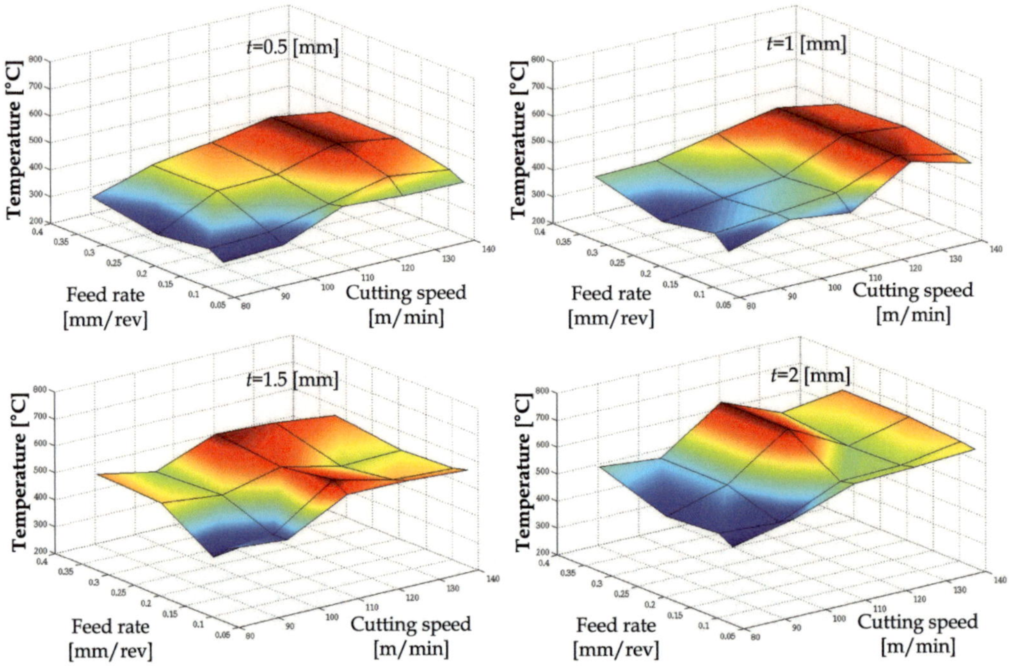

(c)

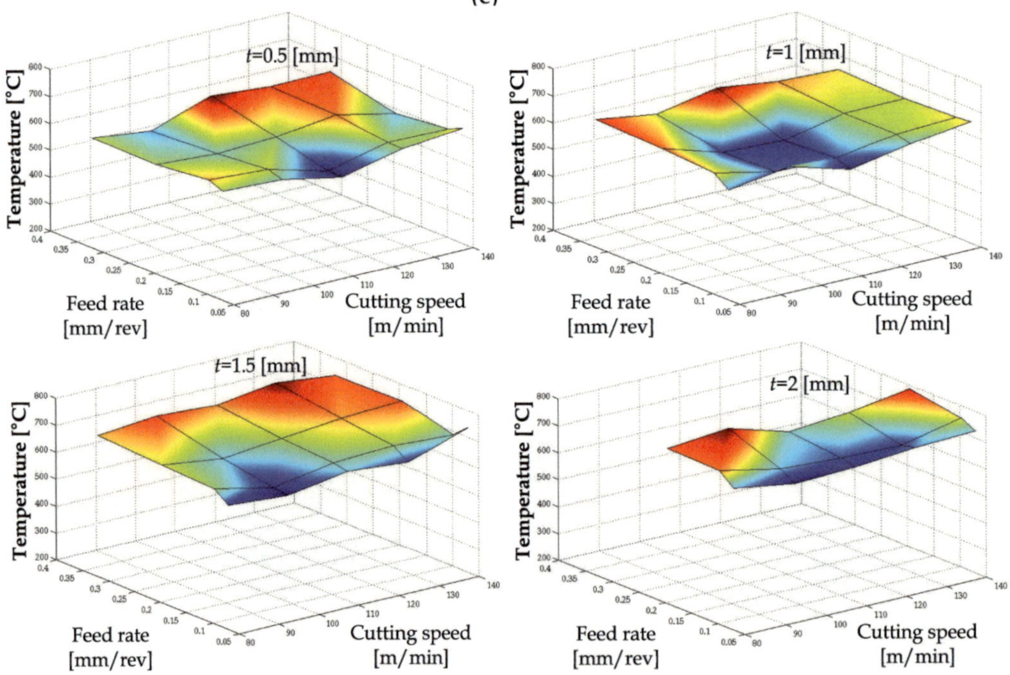

(d)

Figure 5. a) Chip temperature variation for the workpiece with hardness HRC 20; b) Chip temperature variation for the workpiece with hardness HRC 36; c) Chip temperature variation for the workpiece with hardness HRC 43; d) Chip temperature variation for the workpiece with hardness HRC 55

Overall number of experiments carried out is 316, and obtained values can be used for modelling and simulation using various methodologies and FEM analysis. In recent years the research is directed on the use of systems based on artificial intelligence: artificial neural networks, fuzzy logic systems, genetic algorithms, as well as combination of mentioned systems (Tanikić & Marinković, 2011, 2012), (Manić et al., 2005, 2011), (Devedžić et al., 2010), (Tanikić et al., 2008, 2009b).

4. Modelling of the cutting temperature

In this section, results obtained in the first phase are used for modelling of the cutting temperature using response surface methodology, feed forward artificial neural networks, radial basis function network, generalized regression neural network and adaptive neuro-fuzzy system. A comparative study of proposed models is given, and testing of the models was performed on the set of measured data which was not used in the modelling phase.

4.1. Modelling using Response Surface Methodology (RSM model)

Response Surface Methodology (RSM) is a tool for understanding the quantitative relationship between multiple input variables and one output variable. It is the process of adjusting predictor variables to move the response in a desired direction and, iteratively, to an optimum. RSM model is formulated as following polynomial function (Erzurumlu & Oktem, 2007):

$$f = a_0 + \sum_{i=1}^{n} a_i x_i + \sum_{i=1}^{n}\sum_{j=1}^{n} a_{ij} x_i x_j + \dots + \varepsilon \quad i < j \tag{4}$$

where: a_0, a_i, a_{ij}... tuning parameters, n – number of model parameters

Four different models are created from the set of 316 measured data. First model uses only constant and linear terms (Linear model), second model uses constant, linear and squared terms (Pure quadratic model), third model uses constant, linear and cross product terms (Interactions model) and fourth model uses constant, linear, squared and cross product terms (Full quadratic model). The coefficients of the proposed models are shown in Table 1. This method is simpler than standard nonlinear techniques for determining optimal designs.

The set of 122 measured data (testing set), which was not used in the modelling phase was used for models testing. Some of the calculated temperatures with different RSM models are shown in Fig. 6. (32 data sets obtained during machining workpiece with hardness HRC 20). The conclusion is that Full quadratic model has the best characteristics (the least values of maximum as well as mean error), and that model will be compared with the other models, which will be created in the next sections.

Model	const.	H	v	f	t	Hxv	Hxf	Hxt	vxf	vxt	fxt	H^2	v^2	f^2	t^2
Linear	-74.33	6.472	1.671	227.7	102.0										
Pure quadratic	69.13	-7.825	3.248	252.1	109.7							0.193	-0.007	-45.53	-2.692
Interactions	-344.7	12.60	3.860	742.1	83.73	-0.050	-6.597	0.396	-2.076	0.067	-26.65				
Full quadratic	-190.0	-1.888	5.416	727.9	85.90	-0.051	-5.806	0.519	-2.199	0.048	-8.203	0.193	-0.007	-46.12	-2.711

Table 1. The coefficients of the proposed RSM models

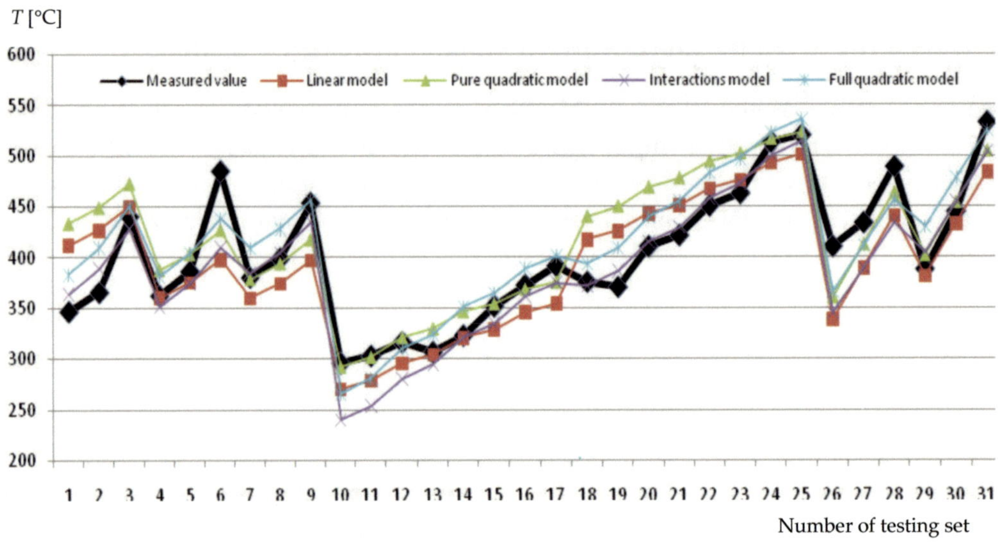

Figure 6. Predicted temperatures with different RSM models

The maximum and mean errors of the proposed models are presented in Table 2.

	Linear model	Pure quadratic model	Interactions model	Full quadratic model
Max. error [%]	23.555	25.247	25.078	18.343
Mean error [%]	8.193	7.218	7.609	6.901

Table 2. The errors of the different RSM models

4.2. Modelling using Artificial Neural Networks (ANN model)

Artificial Neural Network (ANN) is a structure which is able to receive input vector $I=[i_1, i_2, \dots, i_n]$, and generate appropriate output vector $O=[o_1, o_2, \dots, o_m]$ (Santochi & Dini, 1996). The ANN contains several connected elementary calculation units, which are called neurons. Fig. 7. shows a schematic representation of an artificial neuron with input vector (with r elements) and characteristic structure of the feed forward ANN with k hidden layers.

Each of the input elements $x_1, x_2, \dots, x_r$ is multiplied with the corresponding weight of the connection $w_{i,1}, w_{i,2}, \dots, w_{i,r}$. The neuron sums these values and adds a bias b_i (which is not present in all networks). The argument of the function (which is called transfer function) is given as follows:

$$a_i = x_1 w_{i,1} + x_2 w_{i,2} + \dots + x_r w_{i,r} + b_i \quad (5)$$

while neuron produces output:

$$y_i = f(a_i) = f\left(\sum_{j=1}^{r} x_j w_{i,j} + b_i\right) \quad (6)$$

This output represents an input to the neurons of another layer, or an element of the output vector of the ANN.

In this particular case, input layer of all created ANNs has four neurons: (1) Material hardness, (2) Cutting speed, (3) Feed rate and (4) Depth of cut, and only one output neuron for predicting chip's top temperature.

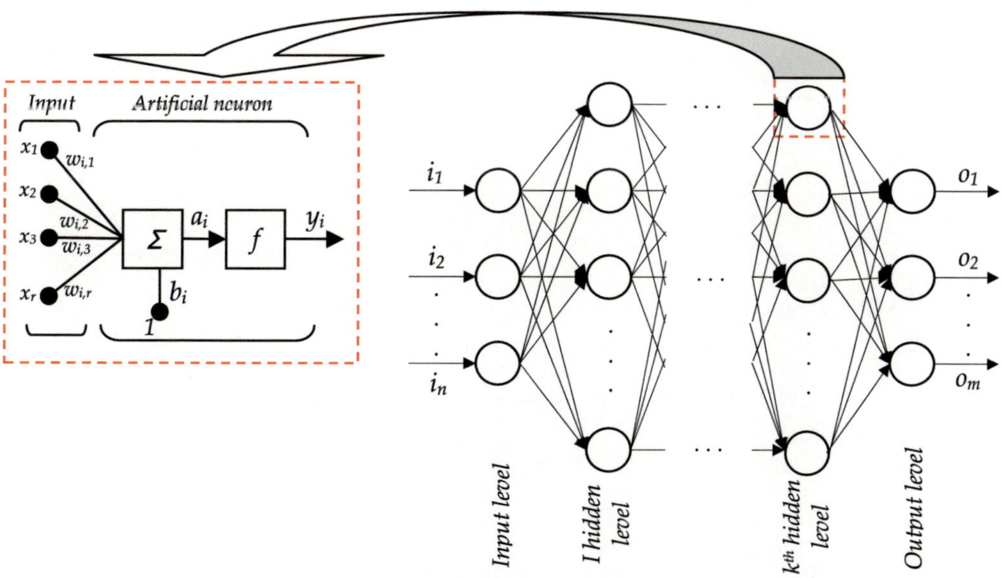

Figure 7. Artificial neuron and the structure of the feed forward artificial neural network

The artificial neural networks, as known, can have random number of layers and random number of neurons in them. Performance of ANN depends on the number of layers, number of neurons, transfer function, presence of a bias as well as on the way the neurons are connected. Unfortunately, there are no formal rules for proper choice of mentioned parameters. In principle, determining all of these parameters is done using personal skills and experience. In the present work six different neural networks with different number of layers and neurons are created in order to achieve minimum error on the one side and avoid overfitting (the situation when the network has a low capability of generalization) on the other side. The structure of the proposed networks are: ANN 4-3-1 (one hidden layer with 3 neurons); ANN 4-5-1 (one hidden layer with 5 neurons); ANN 4-10-1 (one hidden layer with 10 neurons); ANN 4-2-2-1 (two hidden layers with 2 neurons in each layer); ANN 4-5-3-1 (two hidden layers with 5 and 3 neurons, respectively) and ANN 4-10-5-1 (two hidden layers with 10 and 5 neurons, respectively). The main goal is to minimize the performance function, in this case mean squared error (*mse*) function, which can be calculated as:

$$mse = \frac{1}{Q}\sum_{k=1}^{Q} e(k)^2 = \frac{1}{Q}\sum_{k=1}^{Q}\left(t(k) - y(k)\right)^2 \tag{7}$$

where: Q – number of experiments, $e(k)$ – error, $t(k)$ – target values, $y(k)$ – predicted values

The training algorithm used in all cases is Levenberg-Marquardt algorithm which provides the best convergence in the cases of approximation of an unknown function (function prediction), and the number of training cycles is 500. The neurons in input and hidden layers of ANNs have sigmoid transfer function, while the neurons of the output layer have linear transfer function. Minimization of the *mse*, depending on number of training cycles, for various ANNs configurations is shown in Fig. 8.

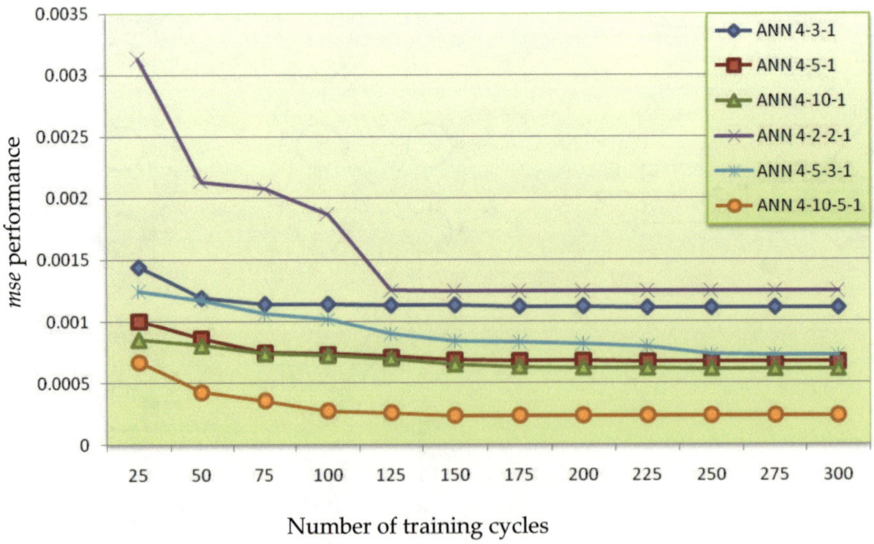

Figure 8. *mse* performance function depending on the number of training cycles

After the training phase, the ANNs were tested. The set of 122 measured data (which was not used in the training phase) was used for testing the ANN models. Maximum and mean error of all proposed networks is given in Table 3.

	ANN 4-3-1	ANN 4-5-1	ANN 4-10-1	ANN 4-2-2-1	ANN 4-5-3-1	ANN 4-10-5-1
Max. error [%]	15.001	14.345	14.461	22.504	24.218	14.236
Mean error [%]	4.387	3.623	3.602	5.615	4.114	3.229

Table 3. The errors of the different ANN models

The ANNs prediction, for the set of 32 measured data (obtained by measuring chip's top temperature while machining workpiece with hardness HRC 20) is shown in the Fig. 9. From given figure and table 3, it can be concluded that ANN 4-10-5-1 shows the best performance, and this network will be used for comparison with other models.

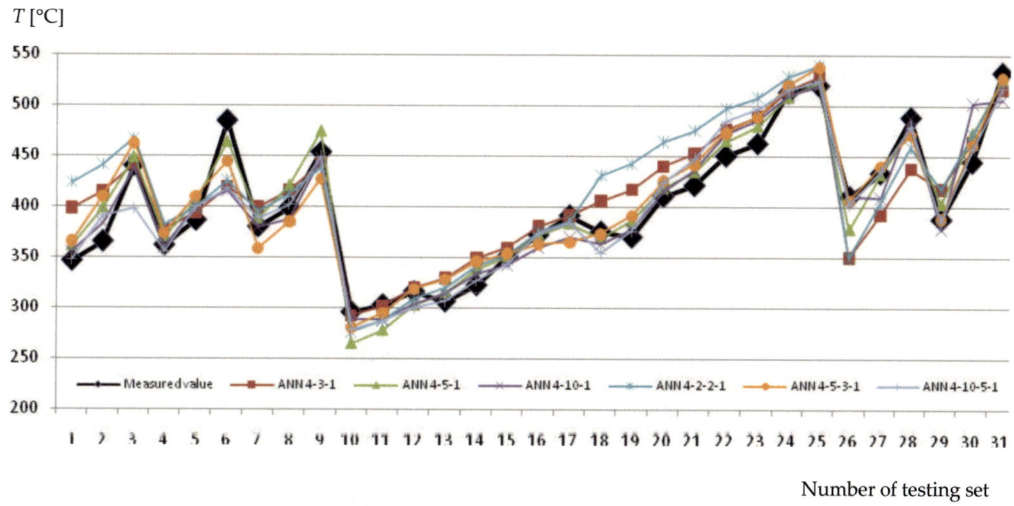

Figure 9. Predicted temperatures with different ANNs architectures

4.3. Modelling using the Radial Basis Function Network (RBFN model)

Radial basis function network (RBFN) employs local receptive fields to perform function mappings (Chen et al., 1991). Fig. 10. shows radial basis neuron and characteristic structure of RBFN. The output of the i-th receptive field unit (hidden unit) is expressed as:

$$a_i = R_i(\vec{n}) = R_i\left(\left\|\vec{x} - \vec{w}_i\right\| b\right), \quad i = 1,2,...,H \tag{8}$$

where: $\vec{x}$ – input vector, $\vec{w}_i$ – weight vector (the same dimensions as $\vec{x}$ vector), b – bias, H – number of receptive field units, $R_i(\cdot)$ – i-th receptive field response with a single maximum at the origin

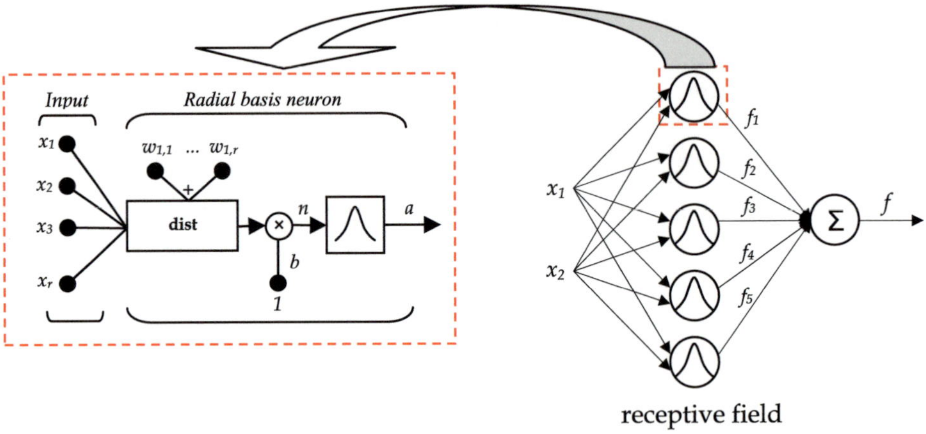

Figure 10. Radial basis neuron and characteristic structure of the radial basis function network

The transfer function for radial basis neuron has output 1 when its input is 0. As the distance between x and w_i decreases, the output increases. The radial basis neurons with weight vectors quite different from the input vector $\vec{x}$ have outputs near zero and these small outputs have only a negligible effect on the overall output. In contrast, a radial basis neuron with a weight vector close to the input vector $\vec{x}$ produces a value near 1. Typical transfer function of radial basis neuron is:

$$R_i(\vec{n}) = e^{-n^2} \tag{9}$$

The output of the radial basis function network can be computed as follows:

$$f(\vec{x}) = \sum_{i=1}^{H} f_i a_i = \sum_{i=1}^{H} f_i R_i(\vec{n}) \tag{10}$$

In this case, set of 316 measured temperature data (and corresponding cutting regimes) was used for creating a radial basis function network, while the rest 122 data (testing set) was used for RBFN model testing. Maximum and mean errors of this model are 44.142% and 8.801% respectively. The prediction of the chip's top temperature of this model is given in the Fig. 11.

4.4. Modelling using the Generalized Regression Neural Network (GRNN model)

A General Regression Neural Network (GRNN) is a special type of neural network with radial basis function which is usually used for function approximation. It has two layers: radial basis layer (which is identical with radial basis layer in RBFN), and second, special linear layer (Wasserman, 1993). The number of neurons in radial basis layer is equal to the number of input/output data pairs. The argument of the radial basis function is a product of the weighed input (distance between the input vector and its weighted vector) and bias b. If a neuron's weight vector is equal to the input vector (transposed), its weighted input will be 0, and its output will be 1.The second layer also has same number of neurons as the number of input/output vectors. Given a sufficient number of hidden neurons, GRNNs can approximate a continuous function to an arbitrary accuracy. Generally, GRNN is slower to operate because it uses more computation than other kinds of networks to do their function approximation, but, taking in consideration the speed of the modern computers, this disadvantage became minor.

The same set of 316 training vectors was used for GRNN modelling, and testing was performed on the set of remaining 122 input/target vectors. The maximum error which GRNN produced is 16.907%, while the mean error is 2.827%. The graphic representation of the prediction of this network is given in the Fig. 11.

4.5. Modelling using hybrid, Neuro-Fuzzy system (NF model)

Adaptive Neuro-Fuzzy (NF) systems represent a specific combination of artificial neural networks and fuzzy logic, so they combine the ability of learning of artificial neural networks with the logical interpretation of fuzzy logic systems (Sick, 2002). The basic rule of the adaptive networks learning is based on a descent gradient method which was proposed in the 70s of the previous century (Werbos, 1974). Adaptive neuro-fuzzy network consists of many layers of nodes (neurons), each of which performs a particular function (node function) on incoming signals as well as a set of parameters pertaining to this node. The type of the function which the node performs may vary from node to node, and the choice of node function depends on overall input-output function that network simulates (Jang, 1993).

This system represents the way for adjusting existential base of rules, using the learning algorithm which is based on the assembly of input-output pairs, used for training. Taking into consideration some constraints, the architecture of the adaptive neuro-fuzzy system (ANFIS – Adaptive-Network-based Fuzzy Inference System) is equivalent to radial basis function networks. The characteristic architecture of the adaptive neuro-fuzzy system is shown in Fig. 12.

For simplicity, suppose that system has only two input values x and y (Level 1), and one output value z (Level 5). In the case shown in Fig. 12., the rule base consists just of two fuzzy IF-THEN rules (Takagi-Sugeno type), as shown in Level 2:

Rule 1: **IF** $\underline{x \text{ is } A_1}$ **AND** $\underline{y \text{ is } B_1}$, **THEN** $\underline{f_1 = p_1 x + q_1 y + r_1}$

Rule 2: **IF** $\underline{x \text{ is } A_2}$ **AND** $\underline{y \text{ is } B_2}$, **THEN** $\underline{f_2 = p_2 x + q_2 y + r_2}$

First part of fuzzy rule (after the **IF** part of the rule) is called premise, while the second part of the rule (after the **THEN** part of the rule) is called consequent. From the ANFIS system architecture it is obvious that for the given values of premise parameters, the output value can be presented as linear combination of the consequent parameters. Mathematically, this can be presented as:

$$\begin{aligned} f &= \frac{w_1}{w_1 + w_2} f_1 + \frac{w_2}{w_1 + w_2} f_2 = \overline{w_1} f_1 + \overline{w_2} f_2 = \\ &= \left(\overline{w_1 x}\right) p_1 + \left(\overline{w_1 y}\right) q_1 + \left(\overline{w_1}\right) r_1 + \left(\overline{w_2 x}\right) p_2 + \left(\overline{w_2 y}\right) q_2 + \left(\overline{w_2}\right) r_2 \end{aligned} \tag{11}$$

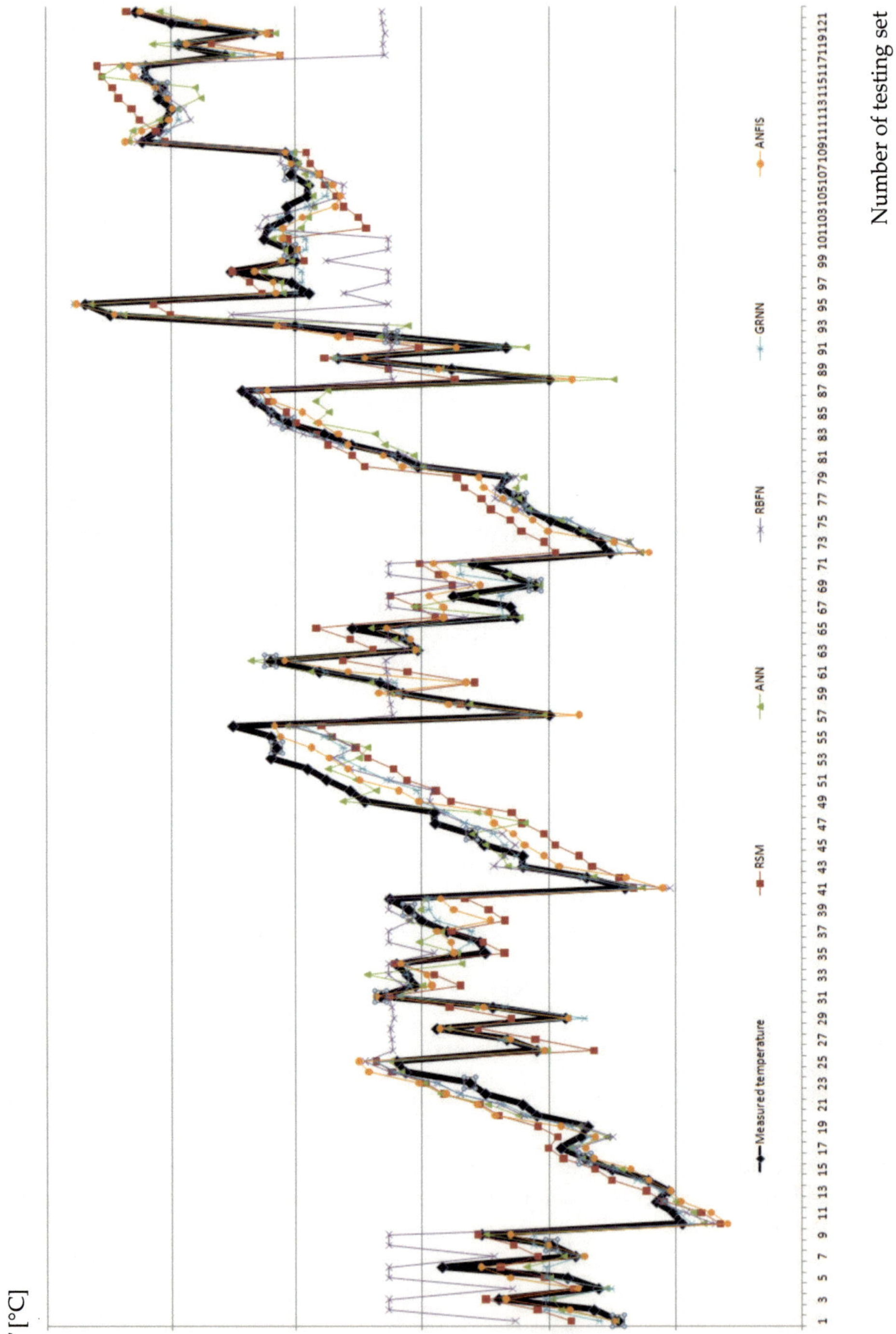

Figure 11. Measured and predicted temperatures with various models

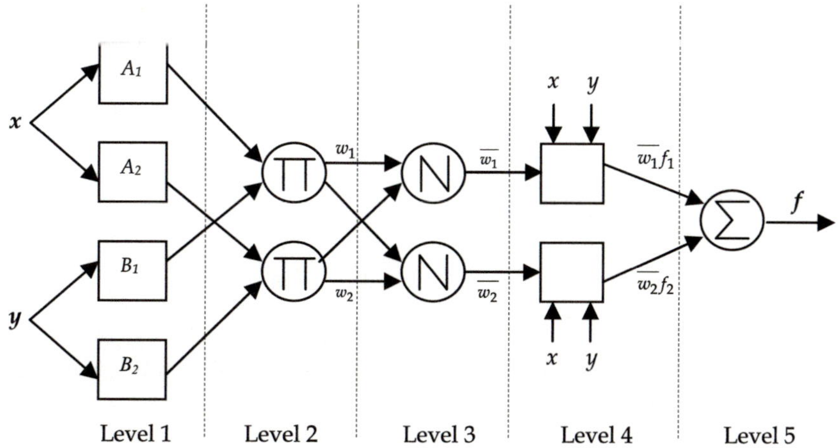

Figure 12. Characteristic structure of the neuro-fuzzy system

The adopted input and output parameters are the same as in other models. All the rules have unity weight, and all output membership functions are of the same type. The number of output membership functions is equal to the number of rules. After the extensive research of the various architectures of the NF systems, and their suitability for the proposed problem, the following parameters are adopted: the number of membership functions of each input is set to 3, the input membership functions are bell shaped, the hybrid optimization method is used, and number of training epochs is 300. A set of 316 measured data is used for training and the rest 122 data sets are used for model testing.

After the successfully finished learning phase, the neuro-fuzzy system accomplished data generalization, and in the modelled field, the value of the chip's top temperature can be predicted without any measurement.

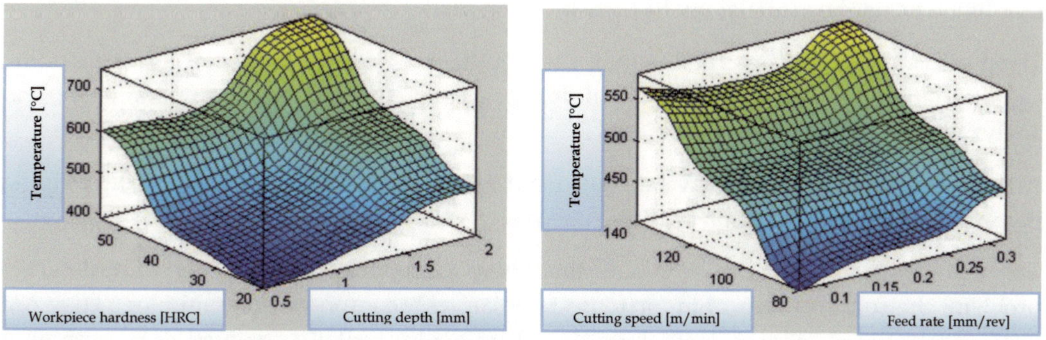

Figure 13. Chip's top temperature depending on two input variables

Graphical representation of the predicted temperature depending on two input variables is shown in Fig. 13 (the input variables combination is arbitrarily). Maximum error achieved in the model testing is 13.439%, while the mean error is 4.319%. The results of the tentative work of the NF system is shown in Fig. 11.

5. Results and comparison of the proposed models

This section presents a comparative study of the proposed models. Testing set of 122 input/target (measured) vectors, not used in the modelling phase was used for checking the accuracy of created models.

The percentage error is defined as follows:

$$e(k)\,[\%] = \left| \frac{t(k) - y(k)}{t(k)} \right| \times 100 \tag{12}$$

where: $e(k)$ – error of the k-th experiment, $t(k)$ – measured value of the k-th experiment and $y(k)$ – predicted value of the k-th experiment

Maximum and mean errors can be calculated as follows:

$$\text{Max. error } [\%] = \max e(k),\ k = 1,2,\ldots,n \tag{13}$$

$$\text{Mean error } [\%] = \frac{\sum_{k=1}^{n} e(k)}{n} \tag{14}$$

where: n – total number of experiments

Response Surface Methodology includes experimental investigations, mathematical methods and statistical analysis.

The advantage of this methodology is in the fact that parameters of the RSM model have meaning, i.e. the coefficient which multiplies some input variable provides some information regarding the influence of that variable on the output value, which is not the case when using the artificial intelligence based models.

The disadvantage of this methodology is the fact that they can be very complex, demanding significant time for data gathering, calculation of all the relevant factors and parameters, and analyzing their influence on the objective.

Artificial neural networks may be considered as parametric functions, and their training involves parameter estimation or fitting process.

The main advantage of the ANNs is that there is no need to explicitly formulate the problem, the solution algorithm, or to be familiar with computer programming. They can also manage noisy or incomplete data, as well as experimentally obtained data, with very complex (or unknown) representation, like in the case presented in this chapter.

The drawback of this methodology is the lack of the principles for determining the optimal network architecture, i.e. the number of layers and the number of neurons in them, as well as the type of their transfer functions. This process can be very complex and time consuming. The principle used in this work implies starting with the minimum number of layers and neurons, creating and training the ANN (tracing the convergence and error), and testing it at the end. After that, the number of neurons and/or layers is raised, and the whole procedure is repeated until the test results became acceptable. As it can be seen, the whole process is highly individual. Finally, the serious drawback of the ANNs is that the resulting performance of the adopted model in an application can't be guaranteed.

Radial basis function networks are successfully used in prediction problems, especially when large number of input variables are present and corresponding output vectors are available, which was exactly the case in our study.

The disadvantage of this model is that they need more neurons and slightly more computational time than standard feed-forward backpropagation networks. Considering the overall operating time of proposed models, this disadvantage becomes irrelevant.

Generalized regression neural network is a special type of RBFN, specialized for function prediction. It can be concluded that maximum and minimum values of Mean errors were obtained using RBFN and GRNN models, respectively. Both networks have the same first, radial basis layer. However, the second, special linear layer (which is present in GRNN model) obviously plays very important role, resulting in the best prediction capabilities among all of the presented models.

Neuro Fuzzy modelling provides a method for the fuzzy modelling procedure to learn information about a data set.

Some of the constraints of the ANFIS are that it must be first or zeroth order Sugeno-type system with a single output, obtained using weighted average defuzzification. Although this can be rectified, in this particular case it was irrelevant since only one output variable (chip temperature) was modelled.

Fig. 11 shows the measured and predicted values of all created models, while Table 4. shows their Maximum and Mean errors. Considering the Maximum error, it is obvious that Adaptive Neuro Fuzzy Inference System has the least value, while Generalized Regression Neural Network is the most accurate when considering the Mean error. The time needed for modelling is very similar for all models. The similar results were obtained when training and testing time were considered. The performances of the present computers make that differences negligible.

All of the proposed models, except RSM model, are black boxes, i.e. the influence of any input variable on the objective is unknown at all, and this fact is one of the main drawbacks of many artificial intelligence techniques.

	RSM	ANN	RBFN	GRNN	ANFIS
Max. error [%]	18.343	14.236	44.142	16.907	13.439
Mean error [%]	6.901	3.229	8.801	2.827	4.319

Table 4. Maximum and mean errors of the created models

The methodologies that are presented in this chapter demonstrate both advantages and disadvantages when compared to each other. The most suitable model for the task under consideration seems to be the Generalized Regression Neural Network and Adaptive Neuro Fuzzy Inference System.

6. Conclusion

The primary goal of this work is examination the possibility of using various models (most of them based on artificial intelligence) in metal cutting temperature modelling. The maximum chip's top temperature was adopted as relevant factor. The infrared method, used in this work, gives a relatively good indication of the measured temperature.

Relationships among the input and corresponding output variables are established from the measured data, as well as trends of temperature changing with cutting regimes and material property changes. Furthermore, modelling of the measured data was performed using the response surface methodology, various types of artificial neural networks and hybrid, neuro-fuzzy system. Almost all of the proposed models can be used for temperature prediction with relatively good accuracy.

Proper selection of the cutting tool, main and auxiliary equipment as well as cutting regimes is of the crucial importance in metal cutting process. Modelling of the main process indicators (such as cutting temperature) can be very useful, and it can help machine shops to machine under optimum conditions, and in that way to reduce the production costs, which is the main goal of any manufacturing production.

Author details

Dejan Tanikić & Vladimir Despotović
University of Belgade, Technical Faculty in Bor, Serbia

Acknowledgment

Research work presented in the paper is funded by the Serbian Ministry of Science within the projects TR34005 and III41017.

7. References

Abukhshim, N.A., Mativenga, P.T. & Sheikh, M.A. (2006). Heat generation and temperature prediction in metal cutting: A review and implications for high speed machining. *International Journal of Machine Tools & Manufacture*, Vol. 46, No. 7-8, pp. 782-800, ISSN 0890-6955

Arshinov, V. & Alekseev, G. (1979). *Metal cutting theory and cutting tool design*, Mir Publishers, Moscow, USSR

Ay, H., Yang, W.-J. (1998). Heat transfer and life of metal cutting tools in turning. *International Journal of Heat and Mass Transfer*, Vol. 41, No. 3, pp. 613-623, ISSN 0017-9310

Boud, F. (2007). Bar diameter as an influencing factor on temperature in turning. *International Journal of Machine Tools & Manufacture*, Vol. 47, No. 2, pp. 223–228, ISSN 0890-6955

Chen, S., Cowan, C.F.N. & Grant, P.M. (1991). Orthogonal Least Squares Learning Algorithm for Radial Basis Function Networks, *IEEE Transactions on Neural Networks*, Vol. 2, No. 2, pp. 302-309, ISSN 1045-9227

Childs, T., Maekawa, K., Obikava, T. & Yamane, Y. (2000). *Metal machining – theory and applications*, Arnold, a member of the Hodder Headline Group, London, Great Britain

Devedžić, G., Manić, M., Tanikić, D., Ivanović, L. & Mirić, N. (2010). Conceptual Framework for NPN Logic Based Decision Analysis, *Strojniški vestnik - Journal of Mechanical Engineering*, Vol. 56, No. 6, pp. 402-408, ISSN 0039-2480

Erzurumlu, T. & Oktem, H. (2007). Comparison of response surface model with neural network in determining the surface quality of moulded parts. *Materials and Design*, Vol. 28, No. 2, pp. 459-465, ISSN 0261-3069

Jang, J.S.R. (1993). ANFIS: Adaptive-Network-Based Fuzzy Inference System. *IEEE Transactions on Systems, Man and Cybernetics*, Vol. 23, No. 3, pp. 665-685, ISSN 1083-4427

Kitagawa, T., Kubo, A. & Maekawa, K. (1997). Temperature and wear of cutting tools in high speed machining of Inconel 718 and Ti-6Al-6V-2Sn. *Wear*, Vol. 202, No. 2, pp. 142–148, ISSN 0043-1648

Kwon, P., Schiemann, T., Kountanya, R. (2001). An inverse scheme to measure steady-state tool-chip interface temperatures using an infrared camera. *International Journal of Machine Tools & Manufacture*, Vol. 41, No. 7, pp. 1015-1030, ISSN 0890-6955

Manić, M., Tanikić, D. & Nikolić V. (2005). Determination of the Cutting Forces in a Face Milling Operation Using the Artificial Neural Networks, *Machine Dynamics Problems*, Vol. 29, No. 3, pp. 51-59, ISSN 0239-7730

Manić, M., Tanikić, D., Stojković, M. & Đenadić D. (2011). Modeling of the Process Parameters using Soft Computing Techniques, *World Academy of Science, Engineering and Technology*, Vol. 59, pp. 1761-1767, ISSN 2010-376X

Marinković, V., Tanikić, D. (2011). Prediction of the average surface roughness in dry turning of cold rolled alloy steel by artificial neural network, *Facta Universitatis, Series: Mechanical Engineering*, Vol. 9, No. 1, pp. 9-20, ISSN 0354-2025

O'sullivan, D. & Cotterell, M. (2001). Temperature measurement in single point turning. *Journal of Materials Processing Technology*, Vol. 118, No. 1-3, pp. 301-308, ISSN 0924-0136

Radovanović, M. (2002). *Tehnologija masinogradnje, obrada materijala rezanjem*, Univerzitet u Nišu, Mašinski fakultet, Niš, Srbija (in serbian)

Santochi, M. & Dini, G. (1996). Use of neural networks in automated selection of technological parameters of cutting tools. *Computer Integrated Manufacturing Systems*, Vol. 9, No. 3, pp. 137-148, ISSN 0951-5240

Sick, B. (2002). On-line and indirect tool wear monitoring in turning with artificial neural networks: a review of more then a decade of research. *Mechanical Systems and Signal Processing*, Vol. 16, No. 4, pp. 487-546, ISSN 0888-3270

Tanikić, D., Manić, M. & Devedžić G. (2008). Chip's temperature modelling using the artificial intelligence methods, *Tehnička dijagnostika/Technical Diagnostics*, Vol. 7, No. 4, pp. 3-11, ISSN 1451-1975 (in Serbian)

Tanikić, D., Manić, M., Radenković, G. & Mančić, D. (2009). Metal Cutting Process Parameters Modeling: An Artificial Intelligence Approach. *Journal of Scientific and Industrial Research*, Vol. 68, No. 6, pp. 530-539, ISSN 0022-4456

Tanikić, D., Manić, M. & Devedžić, G. (2009). Cutting Force Modeling Using the Artificial Intelligence Techniques, *Tehnika/Technics, Mechanical engineering*, Vol. 58, No. 1, pp. 1-6, ISSN 0461-2531 (in Serbian)

Tanikić, D., Manić, M., Devedžić, G. & Stević, Z. (2010). Modelling Metal Cutting Parameters Using Intelligent Techniques. *Strojniški vestnik - Journal of Mechanical Engineering*, Vol. 56, No. 1, pp. 52-62, ISSN 0039-2480

Tanikić, D., Manić, M., Devedžić, G. & Ćojbašić. Ž. (2010). Modelling of the Temperature in the Chip-Forming Zone Using Artificial Intelligence Techniques. *Neural Network World*, Vol. 20, No. 2, pp. 171-187, ISSN 1210-0552

Tanikić, D., Marinković V. (2012). Modelling and Optimization of the Surface Roughness in the Dry Turning of the Cold Rolled Alloyed Steel Using Regression Analysis, *Journal of the Brazilian Society of Mechanical Sciences and Engineering*, accepted for printing, ISSN 1678-5878

Wasserman, P.D. (1993). *Advanced Methods in Neural Computing*, Van Nostrand Reinhold, New York, USA

Werbos, P. (1974). *Beyond regression: New tools for prediction and analysis in the behavioral sciences*, PhD thesis, Harvard University

Wright, P.K. (1978). Correlation of tempering effects with temperature distribution in steel cutting tools. *Journal of Engineering for Industry*, Vol. 100, No. 2, pp. 131–136, ISSN 0022-1817

Young, H.-T. (1996). Cutting temperature responses to flank wear. *Wear*, Vol. 201, No. 1-2, pp. 117-120, ISSN 0043-1648